Fair Use

You should be able to make personal use of your media however, whenever, and wherever you want. Great gizmos like mobile music players and PVRs are making that possible. But Hollywood is locking up the audio and video content you own and taking away your rights. The Electronic Frontier Foundation is fighting back to defend your digital media rights:

http://www.eff.org/IP/fairuse

Member-Supported Since 1990

Make: technology on your time™

Volume 07

Features

Columns

PLANT HACKS
"Everyone wants to put their hand on nature," says Todd Perkins, a flower breeder at Goldsmith Seeds. "It's primal. And there's the joy of taking infinite diversity and selecting just those traits you want." Some of those traits include creating a plant that blooms longer, tastes better, is resistant to disease, or comes in groovy shapes and colors.

PROTO
MIT's Drew Endy follows the path of a fat bumblebee with his index finger. "It's nothing but a flying, reproducing machine," he says. "This object should be editable." Bringing his face inches away from an ant, he says, "Why can't I just hack this stuff?"

ON THE COVER
Florists remove the stamens of lilies because their pollen makes a mess. For the cover shoot, Art Director Kirk von Rohr bought unbloomed lilies with the stamens intact. He immersed the stems in warm water and they blossomed in an hour. Von Rohr says pollen scattered everywhere: "My arms were stained yellow."
Photograph by Howard Cao

Make: Projects

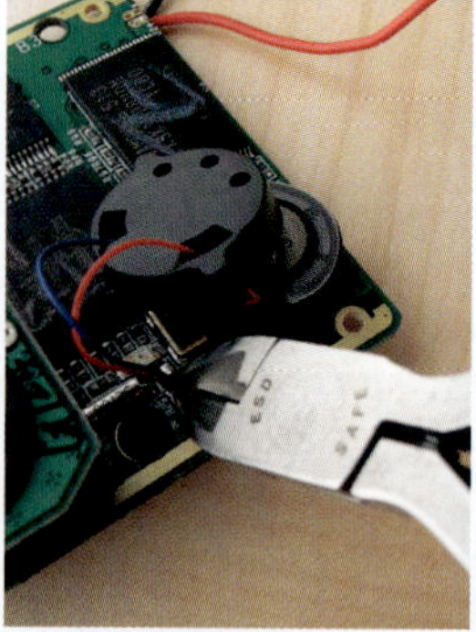

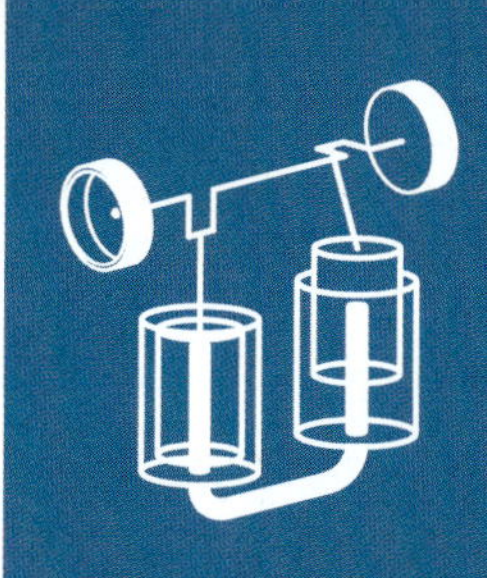

Volume 07

Maker

READ ME: Always check the URL associated with each project before you get started. There may be important updates or corrections.

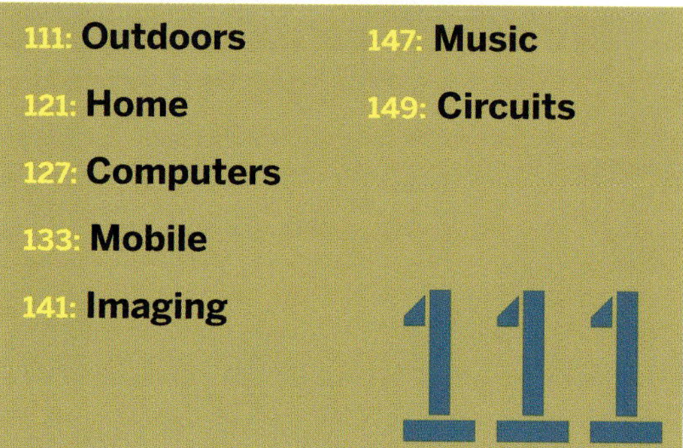

MAKER PROFILE
"The real message of machines isn't that they're helpful workmates," says Survival Research Labs' Mark Pauline. "Like any extension of the human psyche, machines are scary things. When you take the scary human psyche and magnify it hundreds or thousands of times with technology, it's really nightmarish."

Vol. 07, August 2006. MAKE (ISSN 1556-2336) is published quarterly by O'Reilly Media, Inc. in the months of February, May, August, and November. O'Reilly Media is located at 1005 Gravenstein Hwy. North, Sebastopol, CA 95472, (707) 827-7000. SUBSCRIPTIONS: Send all subscription requests to MAKE, P.O. Box 17046, North Hollywood, CA 91615-9588 or subscribe online at makezine.com/offer or via phone at (866) 289-8847 (U.S. and Canada), all other countries call (818) 487-2037. Subscriptions are available for $34.95 for 1 year (4 quarterly issues) in the United States; in Canada: $39.95 USD; all other countries: $49.95 USD. Application to Mail at Periodicals Postage Rates is Pending at Sebastopol, CA, and at additional mailing offices. POSTMASTER: Send address changes to MAKE, P.O. Box 17046, North Hollywood, CA 91615-9588.

It's not taboo!

Crack the case and fiddle to your heart's content...

Learn how to:

- » *Get your current PC ready for an upgrade*
- » *Find the sweet spot for affordable upgrades*
- » *Maintain your system and protect your data*
- » *Replace failed components*
- » *Diagnose hardware problems*
- » *Upgrade your sound and graphics without breaking the bank*
- » *Set up and secure a wireless network*
- » *Keep your hardware cool and avoid overheating*

...and more

O'REILLY®

REPAIRING & UPGRADING YOUR PC

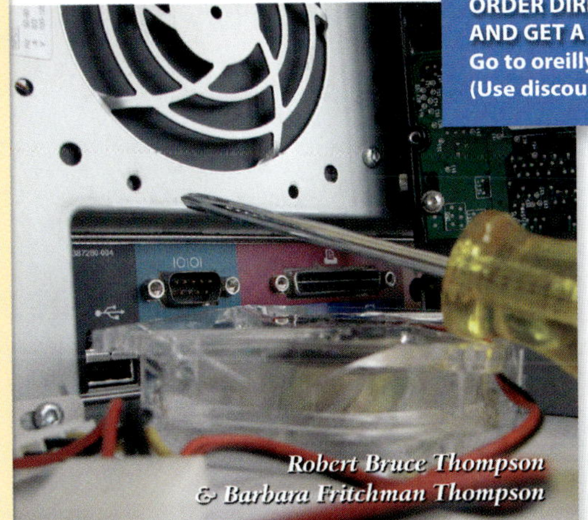

ORDER DIRECT FROM O'REILLY AND GET A 30% DISCOUNT.
Go to oreilly.com/go/repairpc
(Use discount code D6RUPC)

ISBN 0-596-00866-X, 447 pages, $34.99 US / $48.99 CAN

Most computer users think that fiddling with the insides of their PC is taboo. Even those who know how to pop the hood and poke around are often stopped by analysis paralysis—what's causing the system to freeze, which upgrades are the most cost effective, will there be compatibility issues?

Repairing & Upgrading Your PC will get you and your screwdriver unstuck fast. Learn from the experts how to troubleshoot a problematic PC, how to identify which components make sense for an upgrade, and how to take it all apart and put it back together again. With this book at your side, you can tear into your PC with confidence and have fun while you're doing it.

Spreading the knowledge of innovators. www.oreilly.com

EDITOR AND PUBLISHER
Dale Dougherty
dale@oreilly.com

EDITOR-IN-CHIEF
Mark Frauenfelder
markf@oreilly.com

MANAGING EDITOR
Shawn Connally
shawn@oreilly.com

SENIOR EDITOR
Phillip Torrone
pt@makezine.com

PROJECTS EDITOR
Paul Spinrad
pspinrad@makezine.com

EDITOR AT LARGE
David Pescovitz

STAFF EDITOR
Arwen O'Reilly

COPY CHIEF
Goli Mohammadi

COPY EDITORS/RESEARCH
Keith Hammond
Matt Hutchinson
Sanders Kleinfeld
Rachel Monaghan

CREATIVE DIRECTOR
David Albertson
david@albertsondesign.com

ART DIRECTOR
Kirk von Rohr

DESIGNER
Sarah Hart

PRODUCTION
Gerry Arrington

ASSOCIATE PUBLISHER
Dan Woods
dan@oreilly.com

CIRCULATION DIRECTOR
Heather Harmon

ADVERTISING COORDINATOR
Jessica Boyd

SALES & MARKETING ASSOCIATE
Katie Dougherty

MARKETING & EVENTS COORDINATOR
Rob Bullington

ONLINE MANAGER
Terrie Miller

MAKE TECHNICAL ADVISORY BOARD:
Gareth Branwyn, Limor Fried, Joe Grand, Saul Griffith, William Gurstelle, Bunnie Huang, Tom Igoe, Mister Jalopy, Steve Lodefink, Erica Sadun

PUBLISHED BY O'REILLY MEDIA, INC.
Tim O'Reilly, CEO
Laura Baldwin, COO

Visit us online at makezine.com
Comments may be sent to editor@makezine.com

For advertising and sponsorship inquiries, contact:
Dan Woods, 707-827-7068, dan@oreilly.com

Contributing Artists:
Scott Beale, Howard Cao, Roy Doty, Nick Dragotta, Leah Fasten, Tom Giesler, A. Hoekzema, Dustin Amery Hostetler, Jemma Hostetler, Timmy Kucynda, Tim Lillis, Michael Light, Christopher Lucas, Dave McMahon, Jen Munch, David Ng, Nik Schulz, Damien Scogin

Contributing Writers:
Esther Abd-Elmessih, Tim Anderson, Bill Barminski, David Battino, Colin Berry, Joost Bonsen, Keddie Brown, Rob Bullington, Shawn Carlson, Peter Danielson, Cory Doctorow, Nick Dragotta, George Dyson, Russ Ethington, Joanne Fox, Saul Griffith, William Gurstelle, Timothy B. Hewitt, Xeni Jardin, Mister Jalopy, Daniel Joliffe, Peter Kirn, Mike Kuniavsky, Donna Lee, Carlo Longino, Robert Luhn, Dave Mabe, Merlin Mann, John Maushammer, Douglas McDonald, Jon Nakane, David Ng, Danny O'Brien, Will O'Brien, Tim O'Reilly, Ross Orr, Tom Owad, Bob Parks, Charles Platt, Dave Prochnow, Michael H. Pryor, Joseph Pasquini, Philip Ross, Matthew Russell, Matthew Ruschmann, Erica Sadun, Sparkle Labs, Michael Shapiro, Yas Shirazu, Bruce Sterling, Bruce Stewart, Tim Walker, Megan Mansell Williams, Allen Wong, John Young, Lee D. Zlotoff

MAKE is printed on recycled paper with 10% post-consumer waste and is acid-free. Subscriber copies of MAKE, Volume 07 were shipped in recyclable plastic bags.

Please Note: Technology, the laws, and limitations imposed by manufacturers and content owners are constantly changing. Thus, some of the projects described may not work, may be inconsistent with current laws or user agreements, or may damage or adversely affect some equipment.

Your safety is your own responsibility, including proper use of equipment and safety gear, and determining whether you have adequate skill and experience. Power tools, electricity, and other resources used for these projects are dangerous, unless used properly and with adequate precautions, including safety gear. Some illustrative photos do not depict safety precautions or equipment, in order to show the project steps more clearly. These projects are not intended for use by children.

Use of the instructions and suggestions in MAKE is at your own risk. O'Reilly Media, Inc., disclaims all responsibility for any resulting damage, injury, or expense. It is your responsibility to make sure that your activities comply with applicable laws, including copyright.

Interns: Matthew Dalton (engr.), Adrienne Foreman (web), Jake McKenzie (engr.), Ty Nowotny (engr.)

Contributors

Philip Ross (*Home Mycology Lab*) is a San Francisco Bay Area artist who makes sculptural artifacts using living organic materials, including fungus, shellfish, plants, and compost. Through the design and creation of highly controlled environments he manipulates, nurtures, and transforms a variety of living species, much as one might train the growth of a Bonsai tree. In his spare time, he is currently gardening, working on a video about slime mold, and making weird driftwood art.

Hacking is a relaxing hobby, and **John Maushammer** (*Videocam Rocket*) always enjoys figuring out how other people have tackled problems. When the first "single use" digital camera appeared, he couldn't resist the challenge of making it reusable. While not technically a rocket scientist, John has designed satellite electronics. Other hobbies include photography, writing, and exploring new places. His long-term goal is to be able to operate every type of vehicle ever featured in a James Bond movie ... just in case.

Leah Fasten (*Proto* photography) studied economics in college and approaches photo shoots in much the same way she tackled econometric modeling. Each shoot is a giant puzzle. "I'm fascinated by the way environments evolve around the people that inhabit them," she says. "I love what I discover when I'm invited in." Leah lives in Boston with her DNA-replicating scientist husband, her 2-year-old son, a gender-neutral cat, and a geriatric dog. Check out more work at leahfasten.com.

Colin Berry (*Soap Box Derby*) is a journalist, memoirist, fiction writer, and radio commentator. In 2004, his creative work was nominated for a Pushcart Prize. *Spinout* follows an earlier memoir, *Shattered*, published online at onthepage.org. In 2000, Colin and his wife left the hustle and bustle of city life for the serene redwoods of Guerneville, Calif., where he's currently writing a book called *Urban Emigrants*. Colin likes to keep in motion topic-wise, having published pieces on everything from technology to cooking to film reviews. He blogs about the weirdness of country life at colinberry.com.

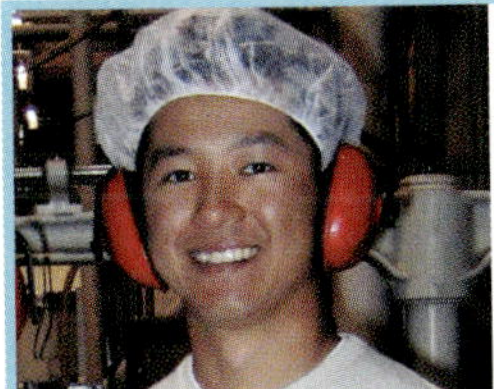

Howard Cao (*Biology section* photography) lives for the three B's: Birdies, Big Macs, and Bigfoot. He's also been known to take a few pictures from time to time. When he's not shooting, Howard can be found golfing on the historic links of San Francisco, researching UFO sightings, racing R/C monster trucks, or whatever ends up becoming his "passion of the month." Visit his tasty goodness at howardcao.com.

Carlo Longino (*Beer Cooler*) lives in Austin, Texas, with his girlfriend and an orange cat named Boo. He generally spends his free time in search of the perfect barbecue, drinking beer, and sweating. His working hours are spent following the mobile telecommunications industry as an analyst for Techdirt Corporate Intelligence, as well as for his own blog, mobhappy.com.

When **Ty Nowotny** (MAKE engineering intern) isn't building cigar box guitars and bottle rockets in the MAKE Lab, he's back in school getting a master's in mechanical engineering. Originally a painter and interactive sculptor, he realized that art and engineering aren't all that different: "Art is still my first love, but we agreed that we should both see other disciplines for a while." He recently reverse-engineered a beloved pair of pants so he can craft them in leather. Anyone have a free leather couch handy?

PLAY HARD, PLAY SOFT

MAKE IS CRAFTING A NEW MAGAZINE.

WE CAN'T SAY NO TO A GOOD IDEA, and our readers give us *lots* of good ideas. You should see how many proposals for great projects come rushing at us through that series of tubes known as the internet, and so many of them are terrific that we're sometimes sorry MAKE isn't a 500-page publication.

So we had a challenge: how to fit more stories in each issue of MAKE without increasing the page count or publication frequency. We thought about printing the stories in a tiny font on rubber paper — to read the magazine, you'd stretch the pages. That idea was nixed. One staffer proposed running the stories through a compression algorithm and printing pseudocode to decompress the text on the inside back cover. That idea was also rejected, and the staffer who proposed it had to stay after work and clean soldering iron tips.

It was time for some out-of-the-box thinking. We took a look at the kinds of project proposals we were getting, and noticed that each story could be placed along a DIY spectrum. There were "soft" projects on one end, and "hard" ones on the other end. For instance, stories about bookbinding, rubber stamping, and silicone molds were on the "soft" side of the spectrum, while making a rocket with a miniature videocamera, a headphone amp in a mint tin, and a wind-powered wi-fi hotspot belonged near the "hard" (not to be confused with "difficult") side.

Ah, we thought — what if we were to collect the best of the soft tech ideas and put them in their own magazine? In other words, give MAKE a sister zine! This publication could showcase the artier, craftier project ideas that came our way. And so, the idea for CRAFT was born.

Just like MAKE, CRAFT is a quarterly, project-based magazine dedicated to the current renaissance in the world of crafts. Our vision with CRAFT is to unite, inspire, inform, and entertain a growing community of highly imaginative and resourceful people who are transforming traditional arts and crafts with unconventional and revolutionary techniques, materials, and tools.

CRAFT is written for and by creative DIY enthusiasts, tech-savvy makers and crafters, students (of all ages), teachers, the intellectually curious, artistically inclined, and environmentally aware. People who derive an innate sense of pleasure in finding unexpected ways to repurpose, remake, and reuse materials, art, technology, and devices in their daily lives. In the premiere issue, CRAFT will show you how to:

» Embroider a skateboard
» Make and program an LED tank top
» Convert an old pair of shoes into chic knitted boots
» Weave with pixels
» Create an iPod cozy with felt-making

When we started telling friends about the new magazine, many of them asked us, "Does this mean that MAKE will never again run an article using yarn, paint, clay, or paper?"

Definitely not. We're as happy with MAKE's editoral mix as our readers seem to be. But if you like the craftier side of MAKE, we hope you'll give CRAFT a try.

The premiere issue goes on sale Oct. 17, 2006, and we anticipate a quick sellout (as was the case with the inaugural issue of MAKE). Avoid crushing disappointment and reserve your copy today by subscribing. As a special offer to MAKE readers, visit craftzine.com and claim an inaugural subscription for yourself or as a gift for the crafter in your life. Do so before Oct. 17, enter promotion code MAKE4CRAFT, and receive a cool CRAFT T-shirt. (And if you refuse, we'll round you up for soldering iron tip-cleaning duty.)

The makers of MAKE are an eclectic bunch, many of whom also enjoy a good craft-tech project.

THE ELECTRIC TANK TOP

By Leah Buechley

5. PRINT IT
Now that you have some practice and a feel for things, let's print the laptop bag. Start with a clean screen.

Since the bag is soft, we need to put something stiff inside the bag to make the printing surface a little harder. In this instance, I used my old cutting mat.

Put some masking tape down on the bag as a guide to help line up your screen. The emulsion is just slightly transparent, and you can see the tape through it. Once it is into position, hold it down, glop

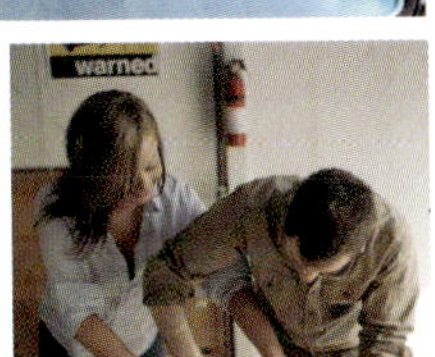

6. CLEAN UP
Place your finished product somewhere to dry (it will take 15 to 30 minutes). Immediately wash your screen — the ink dries fast and can ruin your screen.

You can make about 100-200 prints with your screen. When doing a long run, you may have to periodically wash out your screen between prints to keep the paint from clogging your design.

MAKE IT WASHABLE

If you are printing on a shirt, you need to do another step to make it permanent. Here are two options:
» Iron the shirt on a high, dry setting, placing wax paper between the shirt and the iron.
» Add a few drops of an additive like Versatex Fixer to the paint before you apply it to the shirt. You can mix it right into the ink container.

SARA: I was always a bit hesitant to try silk-screening. It always seemed like I would have to take a semester-long class to learn to print a T-shirt. Kirk's process is something that I can actually do at home, and now I'm not scared to get a little magenta paint under my fingernails, instead of just on top.

KIRK: It really is pretty simple. And once you get the hang of things, you can have a screen ready for printing in an afternoon. I love the immediacy of simple replication that is inherent in silk-screening. Who doesn't love a hand-printed bag, card, or T-shirt? Now go do it!

FINISH

Craft: 113

make cool stuff

O'REILLY craftzine.com

34 Craft: Volume 01

Craft: 35

MAKER'S CORNER

By Dan Woods, Associate Publisher

Advice and news for MAKE readers.

More Cool MAKE Gifts

If you haven't checked out the expanded MAKE online store recently, you're missing some awesome new kits and maker toys, like Joe Grand's Build Your Own Electronic Game Kit; Limor Fried's MiniPOV v2, an inexpensive persistence of vision kit; the MAKE Controller Kit, a modular, programmable controller board specially designed for MAKE by the folks at Making Things; and the ever-popular High-Speed Photography Kit featured in MAKE, Volume 04. Also, while not a kit per se, our hottest product is the TV-B-Gone, a nifty little universal remote device that turns off almost any TV in your proximity. And of course, we also carry back issues, books, collector's editions, and a growing line of cool MAKE T-shirts and sweatshirts.

Stop by and check out the latest merchandise at makezine.com/store.

Encore Maker Faires in the Works

If you were among the fortunate 20,000 makers who converged on the sunny San Mateo County Fairgrounds for the first-ever Maker Faire on April 22nd and 23rd, 2006, our heartfelt thanks for helping to make this event the kind of success that drew attention from CNN, CBS News, The Discovery Channel, and *The Tonight Show*, among others.

If you weren't able to join us, you'll be pleased to know we're working on two new Maker Faires — an encore Bay Area Maker Faire slated for the weekend of May 19th and 20th, 2007, and a Southwestern Maker Faire planned for early fall 2007 in Austin, Texas. And yes, we are also exploring other potential Maker Faire venues across the planet.

If you'd like to be alerted to upcoming MAKE events, including future Maker Faires, visit makezine.com/faire and sign up for the MAKE newsletter.

Be sure to check out page 48 for a special look back at the first Maker Faire.

Auto-Renewal: The Least Expensive Way to Subscribe

Here's the MAKE promise: become a Premier Subscriber, and each year we'll send you a single email notification letting you know when we're about to renew your MAKE subscription and charge your card. You'll always receive a price at least $5 off the standard subscription price. And for being a premier subscriber, you also receive full complimentary access to the entire MAKE archive when you access the MAKE Digital Edition.

To register for the Premier Maker auto-renew service, simply visit makezine.com/premier.

Stay in Touch

Make sure we have your email address on file so you get the latest corrections and updates for projects and features as they come in. Sign up at makezine.com/go/customer; you can choose to receive special offers as well, or corrections only.

Tim O'Reilly

NEWS FROM THE FUTURE

VIRTUAL MEETS REAL.

NOT TOO LONG AGO, PHILLIP TORRONE sent me an interesting bit of news from the future: a "hot or not" application for rating avatars in virtual worlds. And a month or two later, when Second Life (SL) was featured on the cover of *Business Week*, his wife, Beth Goza (who works for Linden Lab, creator of this increasingly popular virtual world), excitedly showed me her avatar's picture in the table of contents: "That's *me*!" she said.

How long before we see people whose self-image in Second Life (or other virtual worlds) is more meaningful to them than their actual physical body? How about a "hot or not" that compares my original self with my imagined self? How about mods to make my physical body look more like my imagined one?

Virtual eventually meets real. We already have a culture of real-world body modification in the form of tattoos, piercings, plastic surgery, steroids, and heck, even regular visits to the gym! Next, we have people adding tech: magnetic sensors wired into fingers, implanted RFID chips, programmable tattoos that can be changed at will. How long before computerized vision augmentation, exoskeletons, and even more advanced technology moves from the realm of medical or military experimentation into the body-mod subculture?

These are small indications that "the presentation of self in everyday life" (to steal the title of Erving Goffman's classic psychology book) is undergoing radical change. Gone is the day when the only identity you could have is the one you were born with. Now, you are who you make yourself to be.

But new selves aren't the only things that people are creating in virtual worlds. They are imagining and making *stuff*. In conventional virtual worlds — massively multiplayer role-playing games — that stuff is somewhat constrained by the storyline of the game: tools, weapons, armor, or whatever. But in freeform worlds like Second Life, where all the vendor provides is raw "land," the economy of stuff is being reinvented from the ground up. Linden Lab has taken the bold step of being clear that anything created by the user belongs to the user — it can be bought or sold. So where other games have underground economies, Second Life's economy is completely upfront.

People are making objects to share or sell: buildings, tools, entertainments, pets ... you name it. As the tools for designing stuff in SL get better, I expect to see more and more things invented and built in SL that don't yet exist in "first life" (aka the real world). We already see the first news from that future: a game called Tringo was invented in Second Life, then sold to a game company for redistribution in first life.

Second Life's currency, the linden, currently trades at about 328 to the dollar (yes, you can convert lindens to real-world currency!), making the current value of the nearly 700 million lindens in circulation about $2 million. There were more than 5 million transactions in April for a total of about $500,000, making the linden perhaps the most widely used micropayments currency on the planet.

And already the hacks are beginning to show the shape of possible futures. Nat Torkington, who still works for O'Reilly here in the U.S. but has moved back to New Zealand, speculated that if he could somehow be paid in lindens, they might be easier and cheaper to convert to New Zealand dollars than U.S. dollars are.

What's more, as long promised by advocates of virtual reality, skills acquired in virtual worlds can well be transferable to real life. *Wired News* recently reported that Yahoo considers a candidate's *World of Warcraft* experience when making hiring decisions. Why? Because to get to high levels in the game, you need the ability to manage groups and lead teams to achieve shared objectives. Meanwhile, net maven Joi Ito recently started a *World of Warcraft* guild that is a hangout for net movers and shakers, noting that "WoW is 'the new golf.'"

Third Life: The state in which it's no longer easy to tell the difference between first life and second life.

Tim O'Reilly (tim.oreilly.com) is founder and CEO of O'Reilly Media, Inc. See what's on the O'Reilly Radar at radar.oreilly.com.

Life Hacks: Overclocking Your Productivity

THE SIMPLEST THING THAT COULD POSSIBLY WORK

GET MORE DONE BY TRYING TO DO LESS.

By Merlin Mann and Danny O'Brien

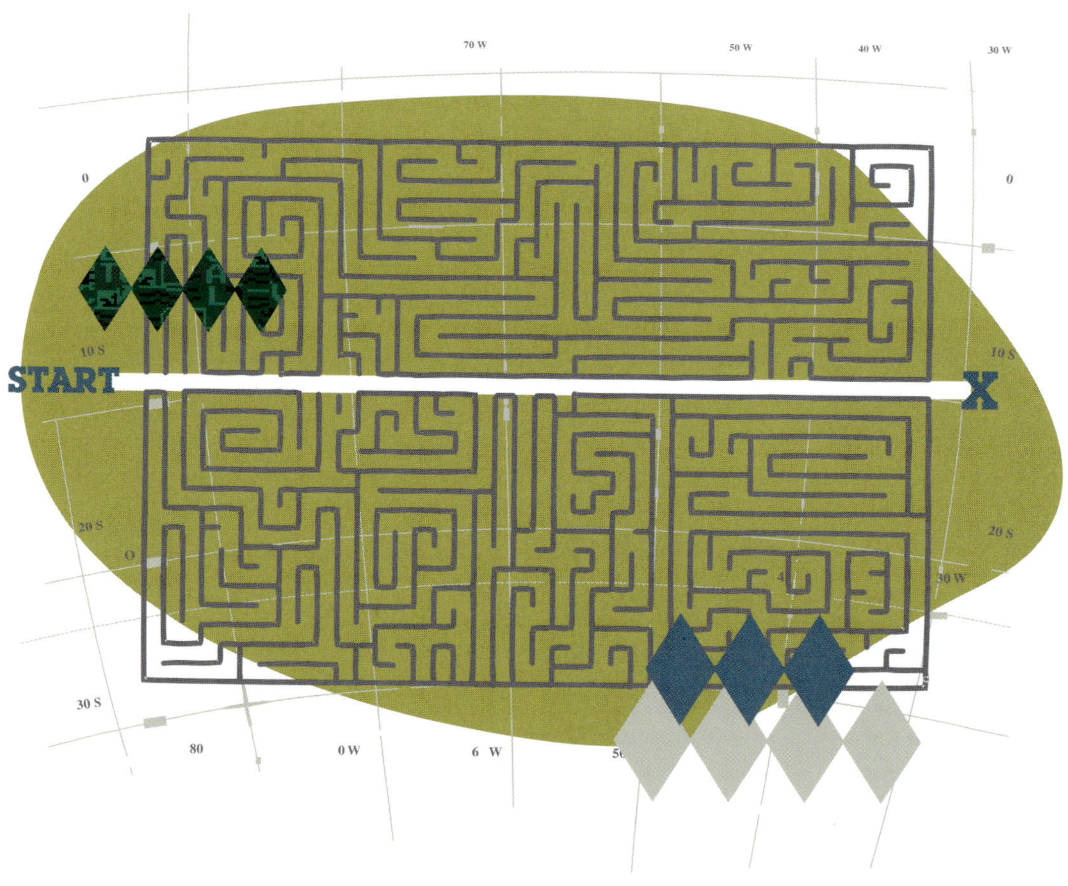

Illustration by Jemma Hostetler

THE BEST LIFE HACKS SHARE MUCH of their conceptual DNA with Extreme Programming (extremeprogramming.org), a software development approach that relies on the principles of teamwork, simplicity, and a baked-in iterative process. (Plus funky names with an increasingly retro feel, but let's ignore that for now.)

Though we'd never want to push the metaphor too far (who, us?), it's fair to say that a Venn diagram of Life Hacks and XP would show a large section of overlap between a reverence for focused simplicity and the pursuit of a clear and modest goal.

Code the Unit Test First

XP programmers gauge the success of a new chunk of code by whether it can pass an automated "unit test." This is a modest programmatic yes/no quiz that determines whether new code returns the kinds of values that would indicate it can play well with all the other pieces of code. Does the `add(x,y)` function return 4 when you feed it 2 and 2? If so, great.

XP has an extra twist (it is extreme, after all), whereby you write this test *before* you begin to develop your code. So you start with a specific goal in mind, and build in the means to determine when you've reached that goal. You know what success looks like, and can recognize when it's OK to move on.

Unit Test? What Unit Test?

Despite (or, more likely, *because of*) this XP/Life Hacks similarity, there's no small amount of irony in the recursive loops of unfocused tinkering that afflict so many aspirational life hackers. When it comes to theoretically improving our productivity, we often do so without the vaguest idea of our own unit test — exactly what problem we're actually trying to fix — let alone the best way to find a satisfying solution and get back to work.

So, we reload del.icio.us all day, download every new web dingus that catches our eye, and then while away our evenings printing out piles of homemade graph paper. The potentially transformative efforts of life hacking are squandered on a wash of amusing time sinks, and you head off to bed 16 hours poorer and no closer to improving *anything*.

So, how do we pass this meta–unit test? How do we stop drifting, quit screwing around, and focus on applying economical solutions to things that actually hang us up? The best answer comes straight out of (where else?) Extreme Programming.

The Simplest Thing That Could Possibly Work

When choosing an approach to building code that will pass their unit test, XP programmers are always encouraged by their beardy masters to "try the simplest thing that could possibly work." Note that this is not the most *comprehensive* thing, nor the most *impressive* thing that could work, nor even a particularly *enjoyable* thing that would just be fun to build.

This approach maps to the results of Danny's original Life Hacks research, where he learned that many über-geeks were overclocking their productivity by creating dozens of ad hoc scripts for use within wildly simplified workflows.

The geeks weren't developing world-beating frameworks that could "scale across the enterprise" or cook French toast every morning — most of the scripts were hastily coded with the single-minded purpose of fixing exactly one problem.

> It's amazing how we can be misled by our ambitions and cravings for novelty.

The geeks' consequent leaps in productivity seemed to come not from simply automating repetitive tasks, but, one imagines, from not blowing two weeks engineering a bloaty system meant to solve every conceivable problem in their lives.

Extreme Life Hacking

The trick is to write down — as in, meticulously outline — what you need to do, and then devise a method that will immediately tell you when your goal has been achieved. Do the least you possibly can to get there. And if it's something you do regularly, try to do even less each time.

It sounds so obvious, but it's amazing how quickly we can be misled by our own ambitions and our cravings for novelty. Extreme Programming works by operating modestly, and by rechanneling the cruel psychological flaws that darken the souls of most geeks and tinkerers. Likewise, if it's not simple and doesn't adhere to a goal with a clear finish line, chances are it's not a particularly good life hack.

Learn how to reel in your mind at Danny O'Brien's lifehacks.com and Merlin Mann's 43folders.com.

Cory Doctorow

A FOR ANYTHING

CREATIVE DESTRUCTION OPENS MORE DOORS THAN IT CLOSES.

IN 1959, MASTER SCIENCE FICTION AUTHOR Damon Knight published a novel called *A for Anything*, a thought experiment about uncontrolled abundance, that happened to predict the internet wars 40 years before they began.

It reads like a story ripped from today's headlines. Chapter 1: A retired banker finds a hand-built invention in his mailbox with a typed note explaining that this is a Gismo and that it will duplicate anything it touches, including other Gismos.

Chapter 2: We meet an FBI agent who's been knocked unconscious by the Gismo's inventor. He comes to and phones the head office. His colleague laughs at him and tells him the world has dissolved into chaos as countless Gismo owners are rioting in the streets.

Chapter 3: We meet the Gismo's inventor. He's holed up with his wife and kids in a mountain cabin outside L.A., watching the craziness he has wrought unfurl below him. A colleague, a hyperkinetic theoretical physicist who always dreamed of being a rocket scientist, finds him in the cabin and talks excitedly of how the Gismo will change space travel: "Put your rocket motors underneath, all you want. With the Gismo, you can have ten or a million. Now what about fuel ... We *make* our fuel as we need it. Forget about your goddamn mass-energy ratios! I can jack up the goddamned Mormon Temple and take it to the moon!"

An ex-Marine shows up at the cabin, commanding a convoy of bomb-rigged cars chained together, each driven by a slave. The Marine has gotten the lay of the land: "Ya gotta have slaves now... [You think] every guy goes off with his own Gismo and that's it? Not on your sweet life, mister. There's just two ways, and you'll find that out — ya gotta own slaves, or ya gotta be a slave." Things go downhill rapidly from here. The Gismo's inventor and his family are enslaved, the physicist is murdered, and by the opening of Chapter 4, we're centuries into the future, a feudal, barbaric society built on Gismo slavery, a dystopia of abundance.

You'll find no better, more prescient parable about the internet wars than *A for Anything*. Today, we're discovering that the internet can make formerly scarce information — everything from how-tos for turning Silly Putty into a trigger for an aerial kite photography rig to, naturally, Hollywood movies, Top 40 music, and popular novels — into abundant information. After all, the internet is a machine for copying bits cheaply, quickly, and without any control.

Now, there are businesses built on scarcity of information — publishing and filmmaking, recording, and even some trades like travel agents and consultants. These businesses have honorable and talented practitioners who bring home the bacon because the information they peddle is scarce. Abundance will put a lot of these people out of business. It's a damned shame.

It's a shame if you're one of the few people who thrived on scarcity. But if you're one of the millions who will benefit from abundance, it's the best news. Ever. If we figured out how to make free hot meals, it would be bad news for restaurateurs, and maybe we'd see less investment in cuisine, but you'd have to be out of your mind to tell the world's starving millions that we're going to try to rid the world of self-replicating dinners to keep the Golden Arches in business.

The dominant narrative of a world of abundance is the dystopia. The music industry and Hollywood want you to believe that infinite, uncontrolled replication will destroy our society, like the foolish Marine in *A for Anything*.

But creative destruction opens more doors than it closes. Abundance needn't mean slaves and barbarism — it can just as easily be a rocket that can take the Mormon Temple to the moon.

Cory Doctorow (craphound.com) is a science fiction novelist, blogger, and technology activist. He is co-editor of the popular weblog Boing Boing (boingboing.net), and a contributor to *Wired*, *Popular Science*, and *The New York Times*.

1+2+3

Penny-Powered LED

By Matthew Ruschmann

Power an LED with some salty water and $1.21.

You will need:
Ice tray, salt, water, pennies (11), dimes (11), paper clips (11), and an LED

To turn pennies into batteries, another electrode and an electrolyte are needed. In this case, dimes (zinc) are used as the positive electrodes and salt water is used as an electrolyte. Such a battery produces a differential of about 0.5V — not enough to light an LED. You need a series of cells, which you can make using an ice cube tray and metal paper clips to hang the pennies and dimes into the electrolyte banks. Because paper clips are conductive, the cells are automatically connected in series to form a more powerful battery, providing a differential of about 2V.

1. Make the solution.

Make a saltwater solution and fill the cells of the ice tray about ¾ full.

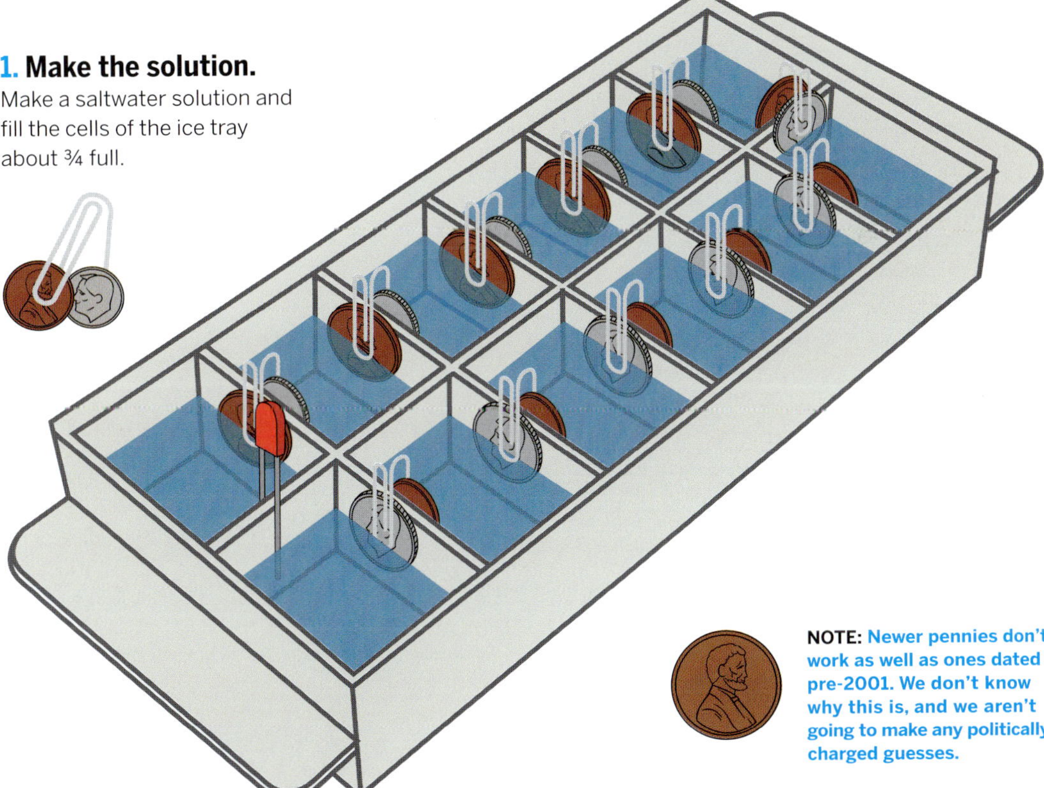

NOTE: **Newer pennies don't work as well as ones dated pre-2001. We don't know why this is, and we aren't going to make any politically charged guesses.**

2. Set up the tray.

Bend each paper clip to hold one penny and one dime, as shown in the diagram. Note that the first pair of cells also has a penny-dime combo clipped to the mid-cell divider, while the last pair of cells doesn't have this. For best results, use a multimeter to confirm a 0.5V reading on each cell first.

3. Light the LED.

To light up the LED, put the short end of the LED into the penny bath and the long end into the dime bath. The penny batteries provide about 110 micro-amps of current in series.

Illustrations by Dave McMahon

Matthew Ruschmann spends his spare time creating power sources that cost less than $2.

Steampunk Superstar

When not digitally animating the creatures he dreams up, **I-Wei Huang** uses steam to breathe life. A 3D animator by day, Huang's extracurricular interests are far from virtual. In his free time, he employs steam engines to power robots and other mechanical motleys that he designs and builds himself — creations fitting snugly into the genre of steampunk (think cyberpunk with Victorian-era technology).

"I loved steampunk even before I knew what it was. I've always had a passion for low-tech machines."

Huang's engines of choice use water vapor at pressures of around 10-20 psi (about what a bike pump can deliver) to move pistons and turn a flywheel, generating power. After buying a miniature steam engine for fun about a year ago, Huang soon yearned for more: a fully functioning, steam-powered, radio-controlled (RC) robot. He searched, but found no such thing. To get a steambot, he'd have to build it himself. Today, Huang's fleet of homemade bots numbers in the dozens and includes a 32-legged steam centipede, an eight-legged steam spider, and a robotic arm called the Armatron.

Huang "Frankensteins" his creatures together using engines, mechanical parts, and toys bought online. Working in his bedroom with little more than a drill and a jigsaw, he hacks the parts to perform to his specifications (walking, crawling, etc.). Since the mini engines have high rpm but little torque, Huang must tweak and re-tweak the gears and pulleys until the parts work as one. Battery-powered radio controls allow Huang to shift the bots between forward and reverse remotely so that, after all his hard work, he can sit back and watch his gadgets go.

Steam may be antiquated, but Huang sees similarities between his low-tech hobby and his high-tech job. "I love seeing things in motion, which is probably why I ended up an animator," he says. For both, "you need a combination of artistic creativity, ability to figure things out quickly, and, of course, attention to how things move and work."

—Megan Mansell Williams

Action videos of Steam Rover at makezine.com/07/made

Steambots: crabfu.com

Photography by I-Wei Huang

First Comes Love

In case you never got around to building an open source solar vehicle (*MAKE, Volume 01, page 44*), get ready for the next generation. What may well be the world's first solar baby carriage was tested early this year, at temperatures far below freezing.

"Useless parking lots can be phased out immediately," announced maker **Jeff Dekzsty**. Easier said than done, but with cars now severely restricted in Central London, and oil prices well on their way up, he may be onto something.

At its preliminary test, the Vee-P1 sped along at 20 kph (about 12 mph, as fast as a bicycle), carrying about 180 lbs., but can be converted to go up to 80 kph (50 mph, not recommended with babies on board). A slim 31 inches wide, it can easily go right into stores, alongside shopping carts and wheelchairs.

"Producing your power *where you use it* has become the premium world directive," says Dekzsty. He should know — his devotion for the past five years to home-built, open source solar vehicles has led him to be called "the solar knight" in some circles. A visit to solarvehicles.org yields food for thought as well as blueprints. One of the FUQ (frequently unasked questions) posted is "Why are all the answers to my questions so complicated and political?"

While he readily acknowledges his debt to other innovators, Dekzsty and his fellow crusaders have been tireless in their pursuit of a vehicle that will be the end of the "Death-Motor State." The solar pram is merely the latest in a long line of vehicles and endless experimentation with various components. "Ah, research," he sighs. "There's really nothing like it under the sun (or moon)."

—Arwen O'Reilly

Solar vehicles: solarvehicles.org

Photograph by Jeff Dekzsty

Eco Roast

Mike and Dave Hartkop had been eyeing their father's abandoned satellite dish in the garage for years. But it took an especially productive night of brainstorming at the local pub to come up with an idea that tapped into their respective interests in coffee and solar energy to put that dish to good use. The Solar Roaster was born.

The first version of the Hartkops' solar coffee-bean roaster was built by attaching several hundred homemade plastic mirrors to the frame of the 10-foot satellite dish. Dubbed Helios, the roaster took about two months to complete and looked more like something off the set of *Battlestar Galactica* than a piece of equipment you'd find behind a coffeehouse.

The Hartkop brothers make a good team for a solar coffee-roasting venture. Mike is the coffee fiend, developing the organic solar-roasted flavors and handling the business details, while Dave is the designer and engineer for their unique roasters. As the solar roasters use no fossil fuels or electricity, Mike likes to claim they've found the most environmentally friendly method of roasting coffee beans in the world.

Photograph by Thomas Hartkop

Dave has been building solar concentrators for the past five years. He's completed two solar coffee roasters and is working on the third and biggest version to date. The solar roasters are getting progressively more efficient, complex, and expensive (they retired Dad's satellite dish after the first version). The hardest part has been building the miniature drum roaster heads, which have to operate with very limited electrical power, handle vibration and wind, and operate when tipped to nearly 90 degrees when tracking the sun.

Helios 3 will be their first mobile solar roaster. The Hartkops plan to take Helios 3 on the road to festivals and shows (and to follow the sun in rainier months).

"I'm hooked on the concept of roasting coffee because the product is instantly accessible by the common person," Dave explains. "It is not an abstract figure given in kilowatt-hours, which supposedly reduces X pounds of fossil-fuel pollution. Solar Roast Coffee reduces those pounds *and* it tastes good!"

—Bruce Stewart

>> **Solar Coffee Roaster:** solarroast.com

Buy My Starship

Trekkies may live for *Star Trek*, but **Tony Alleyne** lives *inside* the fictional space-age world. A science fiction fan since he was a boy, when he grew up Alleyne transformed his suburban British apartment into a meticulous recreation of the starship *Enterprise*.

Alleyne's interior design odyssey began when he fell into a depression following a divorce. A friend tried to console him by loaning him a copy of *Star Trek: The Next Generation Technical Manual*, which documents nerdy construction details of the USS *Enterprise*-D. The more Alleyne read, the more he liked.

"I fell in love with the manual, and finally decided to have a go at building the transporter console," he says.

For Alleyne, who had long worked as a club DJ, the project became a practical journey of self-discovery.

But the distraction that was to transport Alleyne out of his breakup misery took off at warp speed into an all-consuming obsession. After completing a full-scale model of the transporter room, he transformed the bathroom, the hallways, even the kitchen of his condo, and worked on the remodel day and night, seven days a week.

To keep his starship running, he needed money. Eight years and $300,000 in debt later, he was all out, and declared bankruptcy. Word of his Trekified masterpiece spread on the internet. Tabloids, TV, film crews, and fame followed — but paying work doing Star Trek remods did not.

He tried to sell the apartment on eBay for a price he knew was high: $2 million. Nobody bought it, but he's hoping to cash in on newfound fame with a recently completed remodel that updates the apartment to more closely resemble the *Star Trek: Voyager* period.

Doors are rounded like portals; windows are blocked with thick black Plexiglas, to enhance the illusion you're hurtling through space.

Alleyne plans to auction the apartment again, at a more modest price, and he's launched a science fiction interior design firm called 24th Century Interior Design. For a fee, he'll transport himself to your house and transform it into your own private spaceship.

—Xeni Jardin

>> **Star Trek Condo:** 24thcid.com

Photograph courtesy of Tony Alleyne

Burning Down the House

If artist **Aram Bartholl** is afraid of the dark, he couldn't have avoided it any more elegantly. His most recent project, *Random Screen*, is a simple light display that uses old technology (candles and whirligigs) to mimic new technology, toying with our assumptions about the workings of the digital age.

The piece involves a grid of 16 paper boxes, each of which holds a tea light partially obscured by a beer-can pinwheel. The heat rising from the flame spins the pinwheel, and each pixel brightens and dims at a different rate, depending on the intensity of the flame and the speed of the pinwheel. The flickering lights are essentially a random number generator, although in spirit, they're closer to a mechanical and (very) pixilated night sky.

Random Screen was the next generation of a project Bartholl realized in 2004, which was in turn inspired by a 2001 installation in Berlin by the Chaos Computer Club: *Blinkenlights* turned the windows of an office building in Alexanderplatz into a huge, interactive, pixilated screen. Bartholl made his own analog version, *Papierpixel*, using Christmas lights and a 10-foot-long looped punch card to create a simple programmable moving image.

As in *Random Screen*, analog and digital flirt in full view, offering up a creation that approaches history from the back end, the past re-envisioned through the eyes of the future. Both pieces have been shown at a number of galleries around the world, and will be on view at the Ars Electronica festival (themed "Simplicity") this fall.

"From a historical perspective," says Bartholl, "one could speak of the invention of the screen that never existed."

—Arwen O'Reilly

>> **Random Screen:** datenform.de/rscreeneng.html

Stay tuned for the DIY instructions appearing in CRAFT, Volume 01, on sale in October 2006.

Photograph by Aram Bartholl

Time Is on Your (Garage's) Side

If you're driving through the rural town of Panama, N.Y., and forgot to wear a watch, you're in luck. Engineer **John Miktuk** used scrap LEDs and a GPS to assemble a giant clock on the side of his garage.

The timekeeper began as a sign of self-appreciation. Miktuk had a bunch of red LEDs and resistors laying around, scraps from the auto industry, so he drilled holes through the galvanized steel cladding of his four-car garage, plugged them with the lights, built the necessary circuits, and flipped the switch to reveal his surname in glowing letters along a 30-foot wall facing the road.

When it dawned on him that he had plenty of room and materials to add another line of text, Miktuk decided to display something more useful: the time. He circuited together another round of LED-resistor series and connected these to a microcontroller, programmed to ferry information from a GPS unit to the LEDs.

Miktuk mounted a GPS unit 20 feet above the ground and linked it to the microcontroller via a long serial cable. The cable transmits the GPS unit's time and location (calculated from a satellite signal) down the line to the microcontroller, which then directs the appropriate LEDs to turn on. The chip's software even calculates local time from the GPS unit's UTC (Coordinated Universal Time) reading and corrects for daylight saving.

Now anyone with an inkling can build their own LED-GPS clock. In July, Miktuk released "GPS Time on YOUR Garage," a kit for sale on his website for around $300. Although the garage clock is plugged into his home utilities, Miktuk estimates that the entire array (now shining bright green) costs him just $25 a year to power — and it's been running for four years and ticking.

When asked what motivated him to build the clock, he says, "Nothing compares to the sense of accomplishment when a DIY project is finished and working. Except the thrill of the next one. And the next one ..."

—Megan Mansell Williams

» **Garage Clock:** oldvan.com

Photograph by John Miktuk

Photograph courtesy of Busycle

Pedal to the People

If seeing a tandem bike riding by makes you momentarily pause and have a look, imagine your delay when a 15-person bicycle-truck slowly passes by.

Called the Busycle, this bicycle behemoth moves under the power of 28 human legs, plus a driver. And while it can only go about 4 miles per hour, the infectious camaraderie of its passenger power-train just might make it exhilarating.

Boston-based artists **Heather Clark** and **Matthew Mazzotta** had an idea and a lot of energy to see what they could accomplish. Starting with a $1,000 grant from the Berwick Research Institute in Roxbury, Mass., they set to work with limited engineering skills and some borrowed tools.

"We used almost all recycled, reused, and donated materials," says Mazzotta. "Bed frames and weight equipment from Harvard's recycling program to make the frames of the bikes, an old Dodge Van from a moving company as the base of the Busycle, parts from all the bike shops in Boston, and a Mack truck steering wheel from 'The Foot,' a big, bearded guy who runs a tractor-trailer junkyard. We even got some new materials like tube stock and pillow block bearings donated from different Boston businesses."

They posted on Craigslist for anyone interested, and about 65 diverse volunteers joined the effort. "We quickly realized that the talent and support we were seeing in those that wanted to help warranted a much bigger Busycle than we had originally planned on making. And $1,000 to get space [and] buy tools and materials was not going to cut it."

A donation from Sparqs Industrial Arts Club really made a difference; they also helped train volunteers.

On a multi-city tour this past June, the Busycle collected stories, ending each route with a storytelling session. "As for the Busycle's future, we found that the Busycle is not a practical technology," Mazzotta says. "The way we made it on our little budget, it goes too slow. But what it did do was bring a bunch of unique people together, all from different backgrounds, to work together."

—David Albertson

>> **The Busycle:** busycle.com

Bruce Sterling

THE INTERVENTIONISTS

THE "CREATIVE CLASS": TECH GEEKS AND FINE ARTISTS ARE JOSTLING ONTO THE SAME PAGE.

I'M PORING IN WONDERMENT OVER A coffee-table art book. It's *The Interventionists: User's Manual for the Creative Disruption of Everyday Life*. This is one art book with a definite make-do attitude. What maker couldn't want some "creative disruption" in his or her everyday life? Unless that disruption is, like, Jehovah's Witnesses at your door or something.

This art book does not mention Jehovah's Witnesses; basically, it's about ticked-off, way-out Situationist lefties with design, engineering, and art degrees, who use street theater and weird gizmos to freak out and repel global capitalists.

How interesting that a fine art museum has done a popular show about these characters. And the MIT Press did their book, too. How come? I really have to wonder.

It's all about means, motive, and opportunity, that's what. Federal support for the arts has collapsed, so there's no real need for art museums to act all stuffy and official any more. With so little left to lose, they might as well go hog wild. They've got no money left — but these Interventionist artists, just like MAKE types, are noncommercial zealots, so they work really cheap. Their left-wing political antics must be mighty entertaining to the people of Massachusetts, that bluest of blue states. They also tend to work in loose, collaborative, hackerly gangs of activists, which is a remarkable departure for the fine art world. In fact, that makes them look just like the open-source tech world. Huh.

Artists have "audiences." Technologists have "users." This art book insists on calling itself a "User's Manual." It isn't one, because it's got no how-tos, recipes, assembly instructions, or open source code. It's still chatting to a fine art audience of connoisseurs instead of arming the restless masses with weird Yippie gizmos. However, it's getting pretty close.

This book is about artists determined to get right out there and mix it up on the streets. Their museum show sounds way more like a Maker Faire or Bar Camp than it does like some normal weekend museum outing for culture-conscious Boston consumers.

The Interventionists was published in 2004. The museum show in Massachusetts was even earlier. So this art book predates MAKE magazine. Were O'Reilly Media and the Massachusetts Museum of Contemporary Art eagerly comparing notes on the East and West Coasts? No, this is not a conspiracy or a coincidence — this is what we trendspotters call a "groundswell." It's a general cultural change that is manifesting itself in both tech culture and art culture.

Geeks and fine artists are jostling onto the same page. The older labels just don't quite fit in the 21st century. The divisions between art and tech are starting to bore and irritate people. That gap is less relevant to everyday practice. It has less to do with what gets accomplished.

My hunch is that this change has got everything to do with the internet. "Creatives" — that's the new tag — are webbed together now. The art world and the tech world still don't talk to each other exactly — but they can Google each other in a nanosecond. So they're getting cozier, less remote. It's pretty hard to be elitist, obscure, avant-garde, and highfalutin when any smart guy with a browser can slurp up your website and link to all your friends. So why make a big don't-touch-me fuss about your tight little hipster stovepipe? You might as well just spew in all directions and hope for some distributed intelligence.

The art world and the tech world still don't talk to each other exactly — but they can Google each other in a nanosecond.

To the modern powers that be — your city council, for instance — artists, designers, and techies are all "creatives" from the "creative class." That's a mushy label for a genuinely mushy situation. Many of these Interventionists (I'm sure they hate that label) could slide into MAKE magazine with great

Photograph by fi5e

ease. The Institute for Applied Autonomy makes remote-controlled street-graffiti robots. Wow, what maker wouldn't want a few of those? Ruben Ortiz-Torres builds chromed-out low-rider lawn mowers and power tools. They're bitchin', as Steve Jobs likes to say.

Graffiti beam: Katsu's tag on Washington's Arch in New York City's Washington Square Park. Eyebeam's Evan Roth explains how the image is projected from a car: "We just pull over, pop the hood, and run jumper cables through the passenger window to power the projector and laptop."

I'm frankly wondering if I even have to explain the difference between tech and art to anybody under the age of 20. There is a difference, a big one, between doing an art project and attracting curators, collectors, and gallery foot traffic — and starting a Web 2.0 collaborative pop-site, and attracting open source programmers, a horde of users, and Google/Yahoo buyout types. I mean, those are some pretty big differences, right?

But then compare that to the colossal chasm in our society between people who do creative work for love, passion, and intellectual joy, and the profit-driven corporate elite who make all the important decisions. The credentials, the training, the harsh class distinctions between artisans and techies — they just don't seem to matter much anymore. It's no longer "Two Cultures," as C.P. Snow used to call them, at each other's intellectual throats. The 20th century's Two Cultures are being squished together like a Google mashup. Their Cold War is over, even if they don't know it yet. They're becoming the Myrmidons of a digitized culture industry.

The true distinction between art and tech is a simple matter. Techies are dull, geeky people with exciting, practical ideas, while artists are exciting, impractical people with goofy, irrational notions. Frankly, that's more of a slider-bar than a real schism. However, as a "creative class," they've got identical class problems: finding public spaces, finding public attention, and fighting off vampire hordes of commercial IP creeps who want to monetize everything, lock it down with legal barbed wire, and make sure the fix stays in. That's what really matters now. The rest is just tactics.

These sure are exciting times.

Bruce Sterling (bruce@well.com) is a science fiction writer and part-time design professor.

Mark Pauline's Machine Mayhem

For nearly three decades, Survival Research Laboratories has redirected the technology of industry, science, and the military to create the most dangerous theater on Earth.

INTERVIEW BY DAVID PESCOVITZ
PHOTOGRAPHY BY SCOTT BEALE
ILLUSTRATION BY DUSTIN AMERY HOSTETLER

Mark Pauline: "The vision for SRL was always about creepy, scary, violent, and extreme performances that really captured the feeling of machines as living things."

From these scraps monsters are made. Mark Pauline's San Francisco workshop is a jumble of discarded, donated, and questionably acquired "obtanium."

Outfitted with flamethrowers, the radio-controlled, monster-sized walking machines torch piles of broken pianos as a 16-foot Tesla coil spurts crackling blue sparks. A huge air cannon blasts away at plate glass windows before turning ominously, to onlookers who are already suffering from the din of a V-1 jet engine wailing nearby. Legless robo-soldiers crawl from the belly of a two-story Trojan horse as a hovercraft, propelled by four 4-foot-long pulsejet engines, glides chaotically across the burning asphalt. These scenes from a nightmare are real, and this theater of cruelty has a name: Survival Research Laboratories.

Founded in 1978 by Mark Pauline, Survival Research Laboratories is a San Francisco-based network of engineers, artists, hackers, and makers who create "spectacular mechanical performances" where "humans are present only as audience or operators."

"The real message of machines isn't that they're helpful workmates," says Pauline, cleaning the fingernails on his three-fingered right hand. (Part of his hand was blown off in 1982 by a DIY rocket engine. Two of his toes were later attached as replacement digits.)

"Like any extension of the human psyche, machines are scary things," he says. "When you take the scary human psyche and magnify it hundreds or thousands of times with technology, it's really nightmarish."

At an early show, a steel exoskeleton mechanically reanimated a dead rodent while a live guinea pig controlled a large walking machine. At another performance, the giant spring-loaded Hand O' God cocked itself with 8 tons of force before flicking a house of glass to the ground. Meanwhile, the Sparkshooter spewed molten metal 500 yards across the mechanical war zone.

We'll Pay You to Kill!

Right now, however, these mechanical beasts and their brethren are resting inside a dim machine shop where Pauline works in solitude. Wearing oil-stained mechanic's overalls and horn-rimmed glasses, he blends into the mills, drill presses, wire spools, and cartons of unidentifiable raw materials filling every nook. The faint sounds of disco play in the background. Curiosities — doll parts, kitschy posters, vintage prosthetic limbs — hang alongside unusual industrial signs and remnants of surreally comedic props from previous shows.

Pauline's office — until recently his bedroom — overflows with engineering trade magazines, illegible notes, battery chargers, a restaurant-grade espresso machine, piles of work clothes, and cobwebs. A massive poster of dystopian novelist J.G. Ballard looms over a tattered leather sofa. Ballard, who infamously fetishized car wrecks in his novel *Crash*, once defined robotics as "the moral degradation of the machine." Unsurprisingly, Ballard was one of Pauline's main inspirations when he founded SRL in 1978.

Fresh out of art school, Pauline relocated to the San Francisco Bay Area from his native Florida. Immersed in the burgeoning punk scene, he staged prankster attacks on corporate culture and complacency by modifying billboards in clever but acidic ways. In one widely reported stunt, he altered an Army recruitment sign emblazoned with the slogan "We'll pay you to learn a skill!" to read "We'll pay you to kill!" But for Pauline, the infamy was too short-lived to satisfy his calling in life as a creative troublemaker.

"I learned in art school that if you wanted to do something that no one had ever done before, if you wanted to create a truly new idea, you had to be lucky and very dedicated," he says. "Out of that challenge came SRL."

In November of 1978, Pauline, disillusioned with guerrilla billboard art, realized that "the techniques, tools, and tenets of industry, science,

and the military" could be redirected to point out what he saw as the "irony and hypocrisy in the world." So he borrowed the name Survival Research Laboratories from an advertisement in an old *Soldier of Fortune* magazine and produced the first show, a commentary on the oil crisis of the day, in a gas station parking lot. Titled *Machine Sex*, it involved a conveyor belt, a spinning blade, and quite a few (already) dead pigeons.

"The vision for SRL was always about creepy, scary, violent, and extreme performances that really captured the feeling of machines as living things," he says.

Only Marginally Acceptable

The small audience of local punks was impressed and delighted. But most importantly, Pauline had found his fresh idea. He wouldn't learn until years later about kinetic sculptor Jean Tinguely, who constructed an elaborate machine that destroyed itself in the garden outside New York's Museum of Modern Art in 1960.

Three decades later, SRL would stage its own performance at the groundbreaking ceremony for the San Francisco Museum of Modern Art. Following the show, one citizen wrote in a letter to the editor of the *San Francisco Chronicle*, "In Florida, we call it fraud, not art, and we put them in jail." Good thing that Pauline had moved out of Florida long before.

Of course, Pauline knew even before he first put drill to steel for *Machine Sex* that SRL's activities "would only be marginally acceptable to people who wanted to live in peace in urban settings." Indeed, some of what he planned might even be downright illegal. That's why SRL is a legit, tax-paying company.

"I understood that companies could get away with things that individuals can't," he says.

Pathological Amusement

Still, SRL has earned its reputation as a band of troublemakers. In 1989, the group made news after taking credit for a number of mysterious TNT charges that had been found throughout San Francisco. The explosives were fake, grabbed by audience members as unusual souvenirs after a performance and then littered around the city.

In 1995, after SRL's *Crimewave* show at the foot of the Bay Bridge, Pauline was interviewed in connection with the Unabomber case. Although that matter was cleared up quickly, Pauline and a colleague were arrested and charged with using explosives and starting a fire unlawfully.

It's these and a host of other run-ins with fire departments that have made it nearly impossible for Pauline to stage a performance in San Francisco. In recent years, though, the group has packed up flatbed trucks with dozens of tons of equipment and performed in Austin, Los Angeles, Phoenix, and other cities in the United States and abroad. In the 1980s, SRL toured Europe several times with the support of politically well-connected art promoters. These promoters arranged for SRL to have almost unlimited access to scrap yards, and they squelched any potentially threatening controversies immediately.

"They're kind of like the art mafia," he says. In 1999, Pauline and several dozen SRL crew members packed boat-bound shipping containers for the group's first large performance in Tokyo. The show, titled *Thoughtfully Regards: The Arbitrary Calculation of Pathological Amusement*, was sponsored by Japanese telecom behemoth NTT and held in a public park.

"I'm a Vulture Capitalist"

This kind of support is essentially unheard of in the United States, Pauline says. Usually, the show time and location must be a closely guarded secret until the very last minute to prevent the authorities from shutting down the event. Unable to sell advance tickets, Pauline must now bankroll the shows himself at costs of tens of thousands of dollars.

"I'm an artist, but I have to live on an executive salary to do what I do," he says.

For most of his adult life, Pauline barely paid rent at the shop, supporting his tool habit by doing welding and specialized fabrication for high-tech firms in the Bay Area.

"Normally, the research labs contract out their freelance work to shops that have all the right paperwork for things like worker's comp," he

SRL's "most dangerous" device, the *Pitching Machine*, debuted at the 1999 Tokyo performance. The trailer-mounted contraption holds a magazine of thirty 2×4s that are spit out by spinning tires at 135 mph — two boards every second. In Tokyo, it was aimed and fired by an anonymous internet user at a nearby art gallery.

says. "But there were people in those labs who also worked at SRL, so they cut the red tape and steered the contracts to me."

In more than 25 years, SRL has harbored hundreds of engineers, physicists, artists, bathroom chemists, gearheads, and hackers who all find joy in the social commentary disguised as mechanical mayhem.

"SRL has always attracted the interests of people who really are on the cutting edge of technology, and they volunteer assistance and materials that we wouldn't otherwise have access to," Pauline says.

Even with friends in the right places, the contract work couldn't cover the increasing cost of the intricate machine designs Pauline had in his mind's eye and scribbled on scraps of paper. At the end of the tech boom, Pauline identified a new market for his technical knowledge. He buys specialty tech gear — server components, tape drives, scanners — on the cheap, tests them, makes any necessary repairs, and then auctions them off on eBay. Essentially, he has learned to identify the treasures in the tech "trash" that companies cast off due to downscaling or planned obsolescence.

"I'm a vulture capitalist," he says.

While most people lost their shirts on the dot-com bust, Pauline paid off his debt, bought a house, and funded several large shows out of his own pocket.

"I might take a $40,000 loss on a show, but it's just a taxable expense for me and it's considered promotion for the company," he says.

The Fish Boy's Dream

Along with enabling Pauline to move out of his bunker bedroom-cum-office, his vulture-capital career has afforded SRL the luxury of upgrading its tool arsenal. Most of the early machines were assembled from parts that were scavenged from junkyards or obtained "surreptitiously," Pauline says. Items taken without, er, proper approval are known around the shop as "obtainium." Custom components were hand-tooled from raw materials — an incredibly time-consuming process, especially when the machines could take years to build. For example, Pauline spent five years on and off (mostly on) bringing the six-legged walking machine to life. Times have changed. At the center of the SRL shop is a shiny new CNC (computer numerical control) milling machine that automatically fabricates parts based on a digital design file.

"Our cycle of production mirrors recent changes in industrial manufacturing," Pauline says. "You can think more about the design of something because the time it takes to go from bare metal to a finished project is much shorter. All the hundreds of hours you'd spend in front of a manual machine making duplicate parts are condensed way down."

The first maniacal machines to roll off the CNC-powered assembly line were a battalion of *Sneaky Soldiers*. The remote-controlled androids contain a battery-powered chain-drive mechanism so powerful that they can pull themselves along on their steel bellies. Ten soldiers debuted in *The Fish Boy's Dream*, a performance held outside a Los Angeles Chinatown art gallery earlier this year. Of those ten, only two will ever walk — or rather, crawl — again.

The *Sneaky Soldiers* will have to wait for Pauline to tend to their injuries. His attention is now on the *Snot Gun*, a large tube outfitted with a gas mixer and ignition plug. The bottom of the tube is filled with a stinky stew of wallpaper paste and bait fish. When ignited, balls of the gooey "snot" are propelled 200 feet out of the end of the tube. "It's like what happens when you cover one nostril and blow really hard," Pauline explains.

The *Snot Gun* and other machines are being overhauled in preparation for SRL's first large-scale Bay Area show in nearly a decade. This month, the group will perform in San Jose as part of the 13th International Symposium for Electronic Arts. This is one of the rare instances where a city government has given its stamp of approval to Pauline's band of maverick machinists. Maybe they don't know what to expect.

"This show will have an apocalyptic theme loosely based on Dante's *Inferno*," Pauline says. "Think of it as Six Flags over hell."

Outfitted with a massive vertical jaw, the *Inchworm* can inch back and forth or pull itself forward like a crab, thanks to opposing ratchets on the front wheels that drive them in opposite directions.

The six-legged *Running Machine* trots along at 6 mph, powered by a gas engine that drives a hydromotor, that moves the chain linkages, that turn sprockets to enable the locomotion. The *Running Machine* also boasts a hydraulic manipulator arm for picking apart props.

Constructed entirely from aluminum, the *Hovercraft* is propelled and steered via the remote control of four 4-foot-long pulsejet engines providing 70 lbs. of thrust. Roaring at 150 decibels, SRL calls it "the loudest robot in the world."

Spinout

Was building a Soap Box Derby racer my brother's last best chance at escaping his fate? By Colin Berry

ALL HIS LIFE, MY BROTHER KEVIN WAS plagued with terrible luck. It began when he was a teenager in the early 70s, in Longmont, Colorado — our hometown — and soon became something of a family legend. If the Trojan theater was giving away free tickets to *Planet of the Apes*, the kid in front of Kevin in line would get the last one. If Kevin sold enough newspaper subscriptions to win a clock radio, it was broken when he opened the box. If one of his friends shoplifted a pack of Odd Rods bubblegum cards on the way home from school, Kevin got collared for it. It was a pattern. He weathered it well, half-joking about his luck with his shy, gap-toothed grin, but over time it took a terrible toll.

In shop class, however, Kevin seemed to step out from its shadow. He was adept with tools and proved himself a skilled carpenter at an early age. I was seven years younger and remember marveling at the projects he brought home from junior high school: a varnished gun rack; a Newton's Cradle, with its five suspended steel balls; a sturdy set of bedroom shelves for his Revell models. Looking back, it follows that the noisy, meditative setting of the woodshop appealed to Kevin. It was a place where no one shouted at him and where no electronic parts could mysteriously fail.

In our basement, Dad had a woodshop, too, a flagstone-floored, fluorescent-lit grotto with an oversized plank workbench, barrels of wood scraps, and tools hung on a pegboard. It was here, from 1969 to 1972, that my brother built four Soap Box

Photographs courtesy of Colin Berry

Derby racers. He would start in late winter, when snow still covered the ground outside, transforming a small stack of lumber and paper sacks of hardware into a teen-sized, gravity-propelled vehicle.

Balancing the shell of the car across two sawhorses, he built each the same way: a pine plank floorboard supported several plywood bulkheads, to which he anchored Masonite sides and a top. Each car ran on four red-rimmed Soap Box Derby wheels, controlled by a simple cable steering system and foot-pedal drag brake. Each was painted and then lettered with Kevin's name, number, and sponsor logo (Weicker Moving and Storage). And each one got faster.

I am watching him work: the aroma of fresh sawdust mixes with hot electric motor smells, whiffs of plastic wood, and the rubbery tang of new wheels.

Despite their similarities, it was the fundamentals that were really the only thing Kevin's cars had in common. The first two (green and orange) were simple sit-down models; the next two, painted the same bronze color as Weicker's moving vans, were long, elegant, lie-back cars with a rear headrest that supported my brother's head. His bright blue eyes were barely visible from beneath the white regulation Derby helmet.

Leaning against the doorway, I am watching him work: the aroma of fresh sawdust mixes with hot electric motor smells of the drill and jigsaw, whiffs of plastic wood, and the rubbery tang of new wheels. The radio segues from Jerry Reed's "Amos Moses" into Led Zeppelin's "Immigrant Song":

We come from the land of the ice and snow
From the midnight sun where the hot springs blow.

And the gentle *screek screek* of Kevin's file or keyhole saw drowns out the muffled voices coming from upstairs. He is 14; I am 7. I idolize him, of course, and even though he ignores me as he rasps an axle tree into an aerodynamic shape, he probably secretly enjoys having me down here with him.

After dinner, Dad would look in on him, too, but by then the relationship between the two of them had become tempestuous. With his stoic patience, my father could never — would never — understand Kevin's propensity for frustration and impulsive anger. A frozen bolt or misdrilled hole could send him into a furious rage, sabotaging many days' work in a scary tantrum. Generally, he worked alone.

Kevin's labors were part of a long tradition. Started by Chevrolet in 1934, the All-American Soap Box Derby is known as the "World's Gravity Grand Prix," and in my brother's day, it was open only to boys between 11 and 16 years of age. Its rules were strict. Cars couldn't exceed 80 inches in length or 250 pounds, including the driver, and materials couldn't cost more than $40. Contestants built their own cars; parents could offer advice — Kevin could check with Dad on how to laminate the nosepiece or shim the axles (as I did, seven years later) — but the regs specifically forbade adult intervention. Each boy raced in a local competition (ours was in Boulder, on Lehigh Street Hill, around July 4th), and the local winner went to Akron, Ohio, to compete in the national championship. Akron winners received a trophy, a $7,500 college scholarship, and a white champion's jacket that, if I shut my eyes, I can almost picture Kevin wearing.

Each year, his cars got better. The last — in 1972, when he was 16 — was magnificent. It was a sleek, teardrop-shaped stiletto with meticulously trued wheels, a reverse-hinged rear brake, and a steering system that glided gracefully in its guides. Though his early racers were amateurish and hand-spray-painted, this one was elegant and sophisticated, with steel sleeves for cables, a carpeted floorboard, and four airbrushed coats of copper auto body paint applied inside a newspaper-lined cubicle. Most boys Kevin's age couldn't have built it.

At 16, Kevin was heading toward trouble. He already had his learner's permit and had begun associating with boys who drove *real* cars, smoked Winstons, and hung out in his attic room, laughing and talking and listening to what would someday be called Classic Rock. Sometimes they would disappear for a while in Carl Kleveno's blue Duster,

Kevin Berry, the author's brother, built a new Soap Box Derby racer each year from 1969 to 1972. He lost his final Derby race to a cheater who put an electromagnet in his racer. Page 36: Kevin (left) and the author posing in front of another of his racers.

returning with red-rimmed eyes and smelling smoky and sweet. Though we never said it out loud, my family and I knew the 1972 Derby marked the end of something, and we convinced ourselves it was Kevin's last best chance to win.

Race day dawned dry and cloudless. After Dad and Kevin chocked his car carefully into the Fairlane and Mom and my sisters packed tuna sandwiches and a Thermos of milk into the Chevelle, we caravanned to Boulder, securing a spot near the bottom of Lehigh where we could see the finish line. The mood was playful and competitive, as spectators mixed with the young racers.

In his preliminaries, my brother clocked a better time than anyone — more than two seconds faster than the next contender, Bobby Lange, Jr., a rich kid from Boulder with a shiny fiberglass car and cocky attitude. Kevin won his first few heats easily, a copper blur that shot past the finish line, past the checkered flag, and past his sunburned family, who waved and screamed like demons.

"Come on, Kevin!"

"Go! Go! Goooo!"

"Keviiiiiiiiin!"

And the winner is Kevin Berry, Weicker Moving and Storage!

With nearly 30 boys competing, the double-elimination race seemed to last forever. Still, I remember feeling like it was nothing more than a prelude to Kevin's inevitable victory and, although I wasn't conscious of it, a lifting of his tainted luck.

Sometime after 3 o'clock, after an endless succession of heats, only six cars remained: Kevin's, Lange's, and four others. During one race, as Kevin sped past, I saw something strange happen: just past the finish line, his car pulled suddenly to the left, and, rather than braking normally, swerved and plowed into the hay bales piled at the bottom of the hill. Dad and I dropped our pops and sprinted to him.

Dad got there first. "You all right, Kev?" He sounded worried.

My brother had pulled his helmet off, and his face was sweaty and pale. He was clearly distressed. "I'm OK, but I think the car's messed up," he said. "I'm not sure what happened."

Race officials ran over, pulled bales off the car, and carefully lifted Kevin out. He wasn't hurt, but as they rolled his car away, its rear wheels made a jarring shudder. Something was wrong.

"Look at that!" Kevin moaned, pointing, and Dad and I looked. Freshly splintered wood protruded from the foam rubber padding where the axle met the body. Somehow his brake had failed, and the crash had torqued the car badly out of alignment. That was it. Kevin lost his next race by two car lengths, and half an hour later Bobby Lange was the 1972 Boulder champion. I remember riding home in stony silence.

The story could end there — and in a way it did, at least for Kevin. In August, he bought his first real car, a '61 Buick Special, using money he'd made working at Marcantonio's Pizza on North Main. Bobby Lange won in Akron, too; the Boulder *Daily Camera* printed a picture of him, smiling and waving and wearing the white jacket. Kevin's racer went up on blocks.

We didn't pay much attention at first, but the next year, 1973, Bobby Lange's cousin, Jimmy Gronen, also won the Boulder race, and went on to win Akron as well. Yet the officials noticed a strange lurch as Gronen's car came off the metal starting blocks. They X-rayed it the next day and discovered a powerful electromagnet hidden inside the nose of the car. It was wired to a switch that Gronen's head activated when he applied pressure to his headrest, and gave him a jump off the line.

The scandal rocked the Derby. Gronen was stripped of his title, and his winnings were given to the runner-up. But the real blame fell on Gronen's guardian uncle, ski-boot magnate Robert Lange, Sr., Bobby's father. In legal documents and public statements, the elder Lange took full responsibility for the magnet idea (though not its construction), pointing out with indignation that cheating was endemic to the Derby. At some point, officials asked to X-ray Bobby's 1972 car, too — the car that had beaten my brother's — which the DA found during his investigation to have been built with significant engineering expertise and $10,000 to $20,000 worth of equipment. This was clearly in violation of the rules. Though Derby cars are usually preserved for promotional purposes, Bobby's car was nowhere to be found, and remains so today.

None of this really mattered to Kevin. He was past all that, enmeshed in a teenager's life filled with the cars, cigarettes, beer, and drugs that kids in the mid-70s suddenly had to contend with. Within two years, Kevin had accumulated a reckless-driving citation, a DUI, a trip to the police station, and a long succession of real cars, some of them wrecked. Like the radio signal from an interplanetary probe that passes behind a planet, his bad luck, which had seemed to disappear for a while, was back, loud and clear.

Kevin barely graduated from high school, and took a series of jobs working for heating contractors until his patience wore thin. He didn't build much of anything after that — a shingled camper for his pickup, a metal toolbox for Dad — and didn't seem to have any hobbies. He and I grew distant. His friends seemed to disintegrate into desperation or suicide and, in 1998, he did too, with a .22 pistol in the small, neat bedroom of his trailer on the outskirts of Boulder. He died in January, confessing in his note that he couldn't stand working outside in the cold anymore.

What would have happened to Kevin if things had unfolded differently on that July day in 1972? How much would have changed? What happened to his brake? Why did it fail? And if it hadn't, could he have beaten Bobby Lange, even if — and it's all *if*, of course — Lange was cheating? Questions pile up like January snow, obfuscating any real truths and forcing those of us who knew Kevin to turn over a thousand times in our minds the ways it might have gone better. In a way, we — his family — are most to blame for the way we perpetuated Kevin's bad luck in our stories and expectations, allowing it to poke through even as he tried to build something solid against it. Just once, we could have speculated how that long bronze car might have carried him into something better.

A frozen bolt or misdrilled hole could send him into a furious rage, sabotaging many days' work in a scary tantrum.

Despite the scandal, the Derby has survived, though it's been altered almost beyond recognition. Cars are built with kits now, and boys and girls from ages 8 to 17 compete, rally-style, in three different divisions. The rules for each comprise a massive PDF file, and kits start at $400 — not including wheels, which cost up to $100 a set.

Even in Kevin's day, Soap Box Derby wheels were something special. Every year he was issued a new set, and when the car was ready, balancing on its planks, he would slide the new wheels onto their axles, secure the cotter pins, and give them their first long spin. They would whirl for countless minutes, half an hour sometimes, an extended low hiss, like the sound of a distant crowd cheering. There, in the dusty woodshop, it was a sound my brother and I hoped would last forever.

Colin Berry is a journalist, memoirist, and fiction writer who blogs about life in Guerneville, Calif., at colinberry.com.

SHOPPING CART CHAIR

By Tim Anderson

TURN A SHOPPING CART INTO A COMFORTABLE AND STYLISH WHEELCHAIR.

Visitors to the United States are astounded by our shopping carts, which are the largest in the world. Also amazing is their abundance miles from the stores from which they came. No homeless person will ever want for a shopping cart. The best selection of shopping carts can be found near a "redemption center" where homeless people get paid for recycled soft drink containers.

To turn a shopping cart into a chair, look for a cart that really wants to be a chair. That way you won't have to cut or bend much to release the inner chair.

Let's get started. Pound, twist, and pull out the retaining rods and remove the front bumpers. Save them. You can use one later to cover the front edge of the seat. Now, follow along using the photos on the next page.

1. Plan the cuts you will make in the wire basket. Blue tape marks where the cuts will be made.

Iron origami magic will make nice arms from those weird projections. This particular cart had very high sides, so I thought I'd try a new style of arm bending. I've been making these things for a couple of decades, but there is always something new to try.

I used a Sawzall reciprocatiing saw to cut along the lines, but a hacksaw, bolt cutter, or angle grinder will also work well. Hand tools have a lot of advantages over power tools. They're quiet, so you can have a conversation while using them. They're light and easy to carry. They're

Photography by Jen Munch

less likely to leap at you and tear out gaping chunks of your flesh or fling shrapnel into your eyes.

The right chair design will require very few cuts. The chair is already there. As I mentioned, you're just releasing it.

Clamp a couple of boards to make the crease of the bend happen where you want it. This was a tricky diagonal bend and it took a lot of leverage from the long sticks to make it happen.

2. The finished bend. Bending the metalwork hardens it. It's very difficult to change if you bend it wrong.

3. For the next bend, I made a tool from pipes and bolted it onto the protruding arm piece. Even with all that leverage, it takes some wrestling to make a nice curve. File the burrs off the ends of the cut wires. A crutch handle makes a good file handle.

4. Clamp a couple of boards on the front edge of the chair and bend it down. Another method of finishing the front edge is to carve the inner projections of the bumper so they'll fit over the wire, and slide the retaining rod through the holes.

5. Two different chairs, two different styles. Don't leave them outside though, because they will rust.

THE MORAL PROBLEM

Is it stealing to take one of these carts and permanently alter it? The moral problem is complex. You may need to consult a philosopher.

You could buy a new shopping cart for about $100 from the factory. But would it be wrong to bring more shopping carts into a world overcrowded with unwanted ones?

And what if you tried to return the cart to its rightful owner? The iron and other materials in the cart came from stolen Native lands, as does the land you are standing on. Your philosopher will need some time to work on this problem.

In the meantime let's just say, "It's for the children!" and grab as many as we need.

Kids love a shopping cart chair. They can spend hours of fun racing around with it. It's also good as a camera dolly for making home movies. The camera person sits in the chair. You can get amazingly smooth footage zooming down hallways and through rooms.

Tim Anderson, founder of Z Corp., has a home at mit.edu/robot.

Garage Biotech

For a safer world, Drew Endy wants everyone to engineer life from the ground up.

By Bob Parks
Photography by Leah Fasten

Perched on an outcropping of rock at Dartmouth University, Drew Endy follows the path of a fat bumblebee with his index finger. "It's nothing but a flying, reproducing machine," he says. "This object should be editable." He seems especially irritated by bugs all morning as we walk around in the sunshine. Bringing his face inches away from an ant, he says, "Why can't I just hack this stuff?" If you think this guy's spent a little too much time in the lab, you're forgiven.

Since Endy received his doctorate in engineering at Dartmouth and left New Hampshire eight years ago, he's helped kick off an entirely new branch of genetics, remaking biology from an engineer's perspective. The 35-year-old — in jeans, wire-rim glasses, and a funky T-shirt — is a leading proponent of synthetic biology. Unlike genetic engineering, which typically cuts and pastes DNA from existing organisms, synthetic biology seeks to engineer new strands of genetic material from the ground up.

Returning to Dartmouth has probably reminded Endy both how much — and how little — he's accomplished in the last few years. His students haven't exactly revised the bumblebee.

Endy currently teaches at MIT, where he designs new chunks of DNA that are inserted into one-celled animals. Typically, the synthetic beasts in his class are hacked versions of the *E. coli* bacterium. Students swap the genetic material in and out of the organisms like firmware on a circuit board, and then watch to see how the new code performs. When it works, the bacteria change color, transmit information, or execute simple logic functions.

Last semester, for example, students recreated an experiment involving a bacteria camera. They extracted DNA from a photosensitive algae protein, creating a living surface that would hold images when red light was directed at it. The result? Petri dishes bearing Pittsburgh Steelers logos, skulls and crossbones, and biohazard tattoos. The year before, students created a live culture that emulated an oscillator circuit — a DNA routine that acted as an inverter logic gate, making a phosphorescent protein flash on and off over a five-hour period.

"Biology just isn't that hard," says Endy. "You just have to poke and prod and see what happens." At the moment, the only reason why amateur scientists can't perform the same experiments is the high cost of synthesizing DNA. To create the bacteria camera and flashing *E. coli*, Endy's class sent their hacked code to a commercial gene synthesis company named Blue Heron. The company chopped up individual nucleotides to match the students' computer model, sending back live genes in a test tube. The cost was around $10,000. You generally pay by the base pair, so ordering DNA for more complex animals would be astronomical.

Because of the costs, real synthetic biology breakthroughs currently come from well-funded university labs. The field's greatest hits include a project by Princeton electrical engineering professor Ron Weiss, in which he creates bacteria that organize themselves into a bull's-eye shape. Berkeley chemical engineer Jay Keasling is using funds from the Bill and Melinda Gates Foundation to make one-celled machines that pump out the malaria drug artemisinin. Endy himself co-founded a Cambridge, Mass., synthetic biology startup with $13 million in venture capital to commercialize novel biosensors and engineered cells that produce new medicines.

MIT's Drew Endy designs DNA components that can be inserted into single-celled life-forms to make them execute simple logic functions, produce medically useful substances, or work like sensors.

Endy's career arc has been a little strange. This chatty young man certainly never intended to become a university professor or the founder of a biotech startup. Growing up in Devon, Penn., he spent most of his high school years building skate ramps with friends. At nightfall, they'd hop trains in the steel yards for kicks. College didn't engage him until his junior year, when the civil engineering major took a political theory class and started thinking about how ordinary people can effect large-scale changes. He applied to Dartmouth engineering school. It was then that he started connecting his interests to biology.

"Undergrad, I was doing roofing and construction work to make money for summer school," he recalls. "There's a visceral satisfaction to making a physical object. But the first time I cut and spliced a piece of DNA, I felt the same joy of making something. I was like, 'Holy crap! It works!'" To earn his doctorate in engineering, Endy created the deadly virus T7 from scratch. (More recently, he and lab partners put it into an *E. coli* cell and sure enough it reproduced from bug to bug within the strict confines of its container.)

Degree in hand, Endy left school to hack more living stuff. He started working at an independent lab, where he had access to a $80,000 DNA-synthesizing machine. But after months of grueling work, he and coworkers found that they could only make one strand of DNA. "We didn't have the language and grammar yet," he says.

Endy knew what he had to do. Instead of trying to make new biological machines, he would create shortcuts so that non-biologists could build things faster, cheaper, and easier. He began researching the work of Tom Knight, an MIT professor with the novel idea to treat pieces of interchangeable DNA like Lego bricks. Knight calls these standard genetic components *biobricks*. On the MIT site, Endy built a handy public catalog of all known biobricks, called *The Registry of Standard Biological Parts* (located at parts.mit.edu/registry). Many of the successful projects in the annual International Genetically Engineered Machine student competition start with registered bricks from the database. Someday, Endy hopes that engineers will keep their own samples of biobricks, replicating them in live organisms and trading them for free.

At this early stage, only a few of the 1,000 biobricks in the registry are interchangeable. Sometimes, a DNA biobrick simply kills the cell. Other times, the genetic injection cripples the cell so severely that other bacteria take over the culture, ruining the test. Critics such as Caltech professor Frances Arnold say that standardization of DNA simply isn't possible. Genetic code will always act differently depending on the specific environment.

> "Biology just isn't that hard. You just have to poke and prod and see what happens."

The registry also brings up some of the dangers of synthetic biology. Endy has listed his T7 virus there, raising the question of whether a malicious lab or government could construct it for harmful purposes. Endy's answer is that it's already too late to lock down the code. Many pathogens are free for download. The best course of action is to follow the open source model of the software industry: provide free access to the DNA for biobricks and educate as many amateurs and professionals as possible in the ways of using it and spotting trouble.

That's why Endy's latest campaign is to promote the growth of garage biotech. The world will be a safer place, he reasons, if engineers can only see that biology is simply another substrate to hack.

"People know right away what a '7404 hex inverter' is," he says, "But if you mention 'ribosome binding site' or 'transcription terminator,' they act as if aliens are going to show up at the door!" Endy hopes that, in a few years, biology will be further demystified as just another technology, the price of gene synthesis will be more affordable, and rank amateurs will take on ambitious projects. The bugs and the bees may never be the same again.

Bob Parks is a writer and the author of *Makers: All Kinds of People Making Amazing Things in Garages, Basements, and Backyards*.

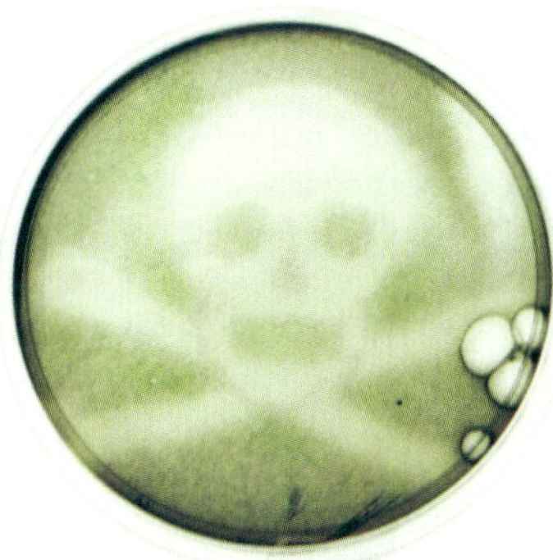

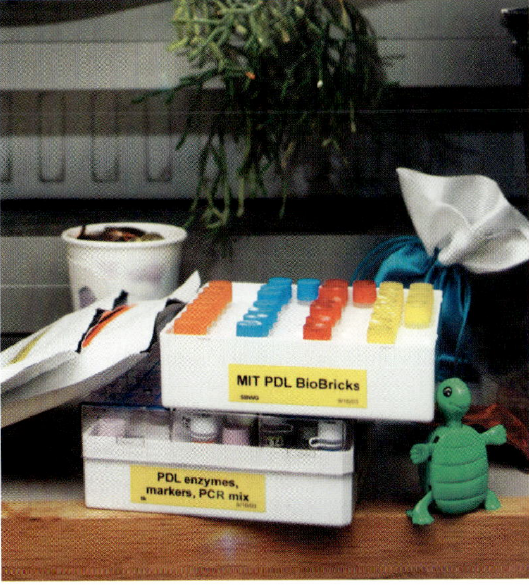

Top: Drew Endy and Natalie Kuldell of MIT's Biological Engineering division prepare a synthetic biology experiment. Above: DNA from a photosensitive algae protein acts as a bacterial camera to develop a skull and crossbones image. Right: Bottles of biobricks — DNA "components" that can be connected like capacitors, resistors, sensors, etc., to form biological circuits

Saul Griffith

HOW TO HOW-TO

USE A HEAD-MOUNTED VIDEO CAMERA TO PRODUCE INSTRUCTIONS FOR MAKING THINGS.

One can now picture a future investigator in his laboratory. His hands are free, and he is not anchored. As he moves about and observes, he photographs and comments. Time is automatically recorded to tie the two records together. If he goes into the field, he may be connected by radio to his recorder. As he ponders over his notes in the evening, he again talks his comments into the record. His typed record, as well as his photographs, may both be in miniature, so that he projects them for examination.
—From "As We May Think" by Vannevar Bush, Atlantic Monthly, *1945*

MY CURRENT OBSESSION IS meta-making, or the documentation side of how-tos. A source of inspiration for my thinking is Vannevar Bush's seminal essay "As We May Think," which has been an influential piece in computer science since it was written in 1945. Most people read it for its incredible foresight as to what the internet would become.

The part I love to latch on to is the "memex," a device Bush describes that links researchers everywhere and the contents of their notebooks so that documentation is seamless, and negative results are registered as often as positive results. It sounds like my fondest dream: it's all wireless, hands-free, and auto-documenting. But that's still the future.

Elegant instructions are quite rare. There are a lot of how-tos out there, in books, on websites, in the instructions for toys and washing machines, even in the seat backs of airplanes, but few are truly great. Lego's original building instructions are good, but those aren't the ones that really count. They're high production value (and high cost) and have been labored over by teams of designers. In a long-tail world, the instructions you'd like to be obvious are your auntie's Roomba hacks, or the Tron guy's instructions on el-wire Halloween costumes. The problem is that documentation is generally so laborious that it is rarely complete or well done.

I listened keenly to the observation by Neil Gershenfeld (of MIT's Media Lab) that the best moment to capture documentation is immediately after someone has just successfully done something for the first time. It's at that moment that people are high on success and wanting to dance around and show the world how they did it. If you can capture that energy toward writing instructions, you'll get the best how-to. The other element that makes instructions top-quality is images, and plenty of them. When inserting a 4/40 left-hand threaded hex bolt into the diaphragm pump housing of your washing machine, no number of images is too many.

I've been thinking a lot about what can be done now to make how-tos a no-brainer. Here's the problem: you want to capture the details as you do it, and probably the first time you do it. Second time around, your hands move fast, and you've figured out the hard bits, so they don't seem so important. The other big problem for me is that I generally use both hands for building things. I often use my mouth too, as two hands aren't enough, and the mouth is such a versatile tool for holding screws and gripping things. Cheek pouches full of pop-rivets — that's my type of squirrel. So with mouth and hands occupied, stopping to take the photograph is damnedly hard.

Will Bosworth models the all-recording eye: A video camera in the backpack is connected by a cable to a bullet camera on the maker's head.

And, I don't know about you, but cameras and greasy welding workshops don't mix well.

My latest solution is a head-mounted video camera. It's been the domain of extreme sportists for a while, but its true home is in the workshop. The solution is still expensive, but I suspect the toy industry will soon solve that problem for us. What I'm experimenting with right now for documentation is pretty simple, but you still have to put it together yourself.

My current solution (as seen above) may not be the most stylish thing you've ever worn on your head, but it doesn't get in your way. And because it's video, you can record the entire project. It's great to have all the footage to choose from.

It's quite simple. I put the camcorder in a Pelican watertight case. The LANC controller and bullet camera emerge from that case, which resides in a slim, comfortable backpack. I've mounted the bullet camera on an old Petzl head-mounted flashlight holder with the controller on the backpack strap.

Unfortunately, editing still takes a long while. iMovie has improved things, but an hour of videotape is still an hour of video to go through, no matter how easy the interface is. I don't yet have any answers to the ultimate in documentation, but I've been learning a lot with the head-mounted camera. I happily recommend it, but what I'd really recommend is for people to explore the solution space until someone figures out the cheapest and simplest way.

There will be a moment in my life when I want to know how to reboot a Prius car computer to run on a higher mix of ethanol using nothing but a Treo, and I hope that I'll be able to get that tutorial when the moment arises. That level of MAKE-ability doesn't exist yet, but I know it has a lot to do with how we document things as a collective. Bring on the memex.

+ Find complete instructions for the Helmet Cam at instructables.com.

YOU WILL NEED:

Sony DCR-HC32 camcorder Good because it has a LANC controller and video input.

Hoyttech bullet camera They make a sweet one with 580 lines of resolution.

Hoyttech LANC controller This is a remote control for the camera.

Pelican case with waterproof feed-throughs (bhphotovideo.com) To keep it all safe from water, oil, dust, and welding sparks.

Saul Griffith thinks about open source hardware while working with the power-nerds at Squid Labs (squid-labs.com).

Photograph by Erich Brandeau

Maker Faire

2006

Genuine Ingenuity

ON EARTH DAY WEEKEND, APRIL 22-23, 2006, more than 20,000 people descended on the San Mateo Fairgrounds in California to experience MAKE's first annual Maker Faire, a combination science fair, craft fair, and celebration of DIY creativity. Over 300 makers showed off their projects, including advanced water rocketry, autonomous robots, homemade chainmail, biodiesel processing systems, potato cannons, wind-powered generators, birdhouses, cameras obscura, steam-powered computers, plywood furniture, RFID implants, fire-breathing trampolines, and much more.

During Maker Faire, members of the media repeatedly asked me the same question: "How do you account for the resurgence of interest in DIY?" My answer was that DIY is fun; it's satisfying and rewarding to make things yourself and share those things with others. There's even a long tradition of DIY technology. And the internet makes it easier to share knowledge, as well as pictures, of what you've created.

However, now that the first Maker Faire is behind us, I've come up with an additional answer. The new interest in DIY is more than just fun; it's part of a deeper search for authentic experiences, something our contemporary culture just doesn't offer enough of. Maker Faire was highly engaging. Unlike so many tech events, there was no one sitting in a corner with a computer checking email or IMing someone. Everyone was fully present, in body and spirit, kids and adults alike.

At Maker Faire, I talked to several youngsters who were carrying things they made at one of the many workshops. One boy, who had made a cardboard crank-toy with a twisted pipe cleaner on top, said, "Look at my satellite." A pair of brothers had made two assemblages out of old laptops and springs; one built a robot and the other made a controller to give orders to the robot. "This talks to that," one of them said. This robot system only worked in their minds, but it really worked, if you know what I mean. If you can imagine it, you can make it.

The true delight of Maker Faire was that you could talk to makers of any age, and see the projects and parts spread out on the workbenches, but you could also see the projects play out in their minds. When you met makers, big or small, you could see the spark in their eyes and you could have a wonderful conversation. Maker Faire brought all kinds of people together, exploring kindred ideas in science, engineering, art, and craft.

An authentic experience is defined by active participation. The process of designing and making something requires a high level of engagement; it requires that you learn to do something and stay engaged until you've accomplished it. When we buy something, we choose to be less engaged, and the experience itself is diminished, just as when we choose to eat fast food rather than cook something for ourselves. When you make something, you create a story, also of your own making, which can be shared easily with others. It's a genuine expression of yourself and of your own ingenuity. It's something you know to be true and something others can trust.

Perhaps what we're seeing is that DIY is a sustained effort to re-make and re-take popular culture. Can we return pop culture to its roots as an authentic form of personal expression? Can we rediscover the creative spirit that exists in everyone?

There are lots of makers, and once you begin to discover them, you realize they are all around you — in your neighborhood or even in your own house. We certainly hope Maker Faire inspired a whole lot of people to see themselves as makers. Because DIY is about more than using tools and making things. It's about creating a culture based on authentic experiences, not manufactured ones.

We hope you'll join us at the next Maker Faire.

—Dale Dougherty

Maker Faire
2006

1

2

3

4

1. "TradeMark" Gunderson shows off his Thimbletron device. 2. A father and son work on a project in the MAKE Play Day area. 3. Observers of the model rocket launch look to the sky. 4. Jamie Hyneman and Grant Imahara of *MythBusters* try their hands at Segway Polo. 5. Alec Bennett's Trampoline Simon was popular with young and old alike. 6. Steve Wozniak helped raise money for EFF by sitting in the Dump Tank. 7. Dan Goldwater of Squid Labs entertains everyone in the MAKE Garage with his LED bicycle. 8. Bruna Palmeria, wearing Diana Eng and Emily Albinski's Inflatable Dress, modeled in Maker Faire's Fashion Show. 9. We had great weather in San Mateo, and everyone enjoyed being outdoors, especially near the Craft midway.

Photography by Scott Beale (1, 3, 4, 5, 6, 9), Kevin Cotter (2, 8,), and James Duncan Davidson (7)

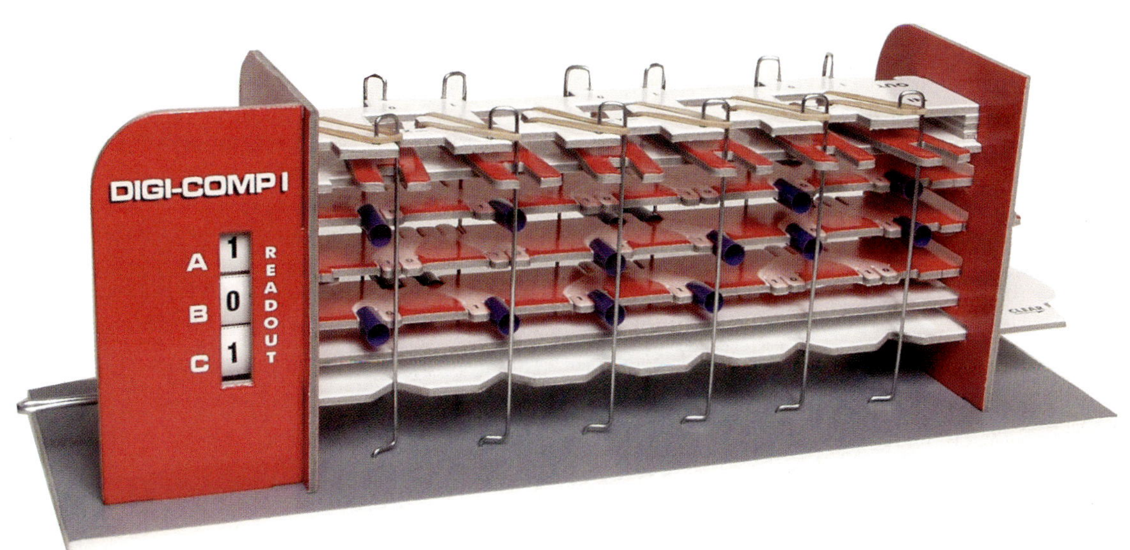

Digi-Comp Redux

A maker in the middle recreates a kit classic.

By Tim Walker

Photograph courtesy Tim Walker

IN EARLY 2005, I STUMBLED UPON A YAHOO group called FriendsOfDigiComp, where several hundred members were reminiscing about the Digi-Comp I, a toy computer from the early 1960s. In messages archived back to 1999, writers sung its praises in posts such as: "I was 11 years old (53 now) when I received a Digi-Comp I for Christmas. I was fascinated with it from a mechanical standpoint .... Even when I had mastered all the programs, I would still get it out and see what it could do. ... I think of it as the spark that got me interested in computing, a career that has been and remains a lot of fun."

The original Digi-Comp — a "visible computer" with a three-bit readout invented by a small New Jersey company called ESR — was one of the most influential educational toys of its era. You could actually program it to do arithmetic, solve logic puzzles, and even play a mean game of Nim. Along the way, it could teach you a whole lot about binary operations and Boolean logic. Priced at about five bucks, some 250,000 Digi-Comps were purchased by parents through Sears and Edmund Scientific catalogs, and by their geeky offspring through innumerable comic book ads.

The idea of remanufacturing the Digi-Comp I in some form was a recurring thread in the Yahoo group. Members posted photos of physical models, hand-sawn from wood or assembled (enormously) from K'Nex pieces. Debates ensued over the relative merits of materials and techniques: laser-cut Lexan, polycarbonate, acrylic plastics, or nylon? Plywood, Masonite, or Honduran mahogany? Sheet metal cut on a milling machine? Injection molding? Duplicates cast from polyurethane or pourable plastic? Quotes were submitted, prototypes alluded to.

One writer enthused, "We need to form a corporation and elect a leader and set up a purser." But in each case, the vision of creating a workable product via email collaboration faded. Six years after the group's formation, the only way to get your hands on a Digi-Comp was to buy one on eBay for about $150.

Forget that. I had to find out how this mechanical computer worked. Fortunately, the group's archives included detailed photos. Several knife blades later, with a somewhat shaky matboard prototype in front of me, it was time for a test drive. At the heart of the mechanism were three horizontal flip-flops attached to the binary readout. Vertical rods established a link between input and output. "Programming" Digi-Comp meant attaching small tubes to I/O tabs along the flip-flops. My simple test routine used only four tubes, which toggled a readout digit between 0 and 1 each time I actuated a clock cycle by moving a small crank.

And the gizmo worked! Yet, even having built it, figuring out exactly how it worked was a challenge. Here was an ingenious toy (technically a "mechanical programmable logic array") that had lost none of its teaching potential. It deserved a new lease on life.

But simply providing a set of plans wasn't the way to go. Building the prototype convinced me that no handheld craft knife could shape all these little tabs, slits, and nubbins precisely enough. It was also way too flimsy; the board thickness would need to double. I began to consider a kit-based approach where I would be the "maker in the middle": fabricating the tricky bits; shipping out a package of subassemblies, loose parts, and instructions; and letting the end user finish the job. I resolved to produce a kit with precision die-cut pieces supplied as prepunched sheets; users could press the parts out and then assemble them.

So began a nonstop, seven-month solo voyage toward "git 'er done." Each component entailed Q-and-A loops, vendor sourcing, and time-and-labor calculations. First, I tested cardboard stock for strength, density, grain, and warp. Eventually, I settled on 80-point binder's board, the stuff used for hardcover books. Next, I had to determine what scale I was going to work at. That would depend on package constraints, which meant planning from the outside in. Online, I discovered gusseted stay-flat mailers: unadorned, practical, 9x12 envelopes. If we die-cut the pieces on a 12"x18" layout, then sliced that down the middle, could we cram sheets, parts, and a 50-page manual into a 9x12 mailer?

Visually, the pieces had to look great, with snap and gloss, legible labeling, and a retro gestalt.

Here was an ingenious toy that had lost none of its teaching potential. It deserved a new lease on life.

DrawPlus, from the U.K.-based company Serif, let me design the vector shapes on one layer and the print wraps on another in registration. Color laser printing had its own pitfalls: not many low-volume print shops handle 12x18 jobs; fewer still can guarantee a flawless cherry red on glossy stock! And there was still the matter of gluing the wraps to the boards prior to die-cutting.

One recurring question was whether to outsource a component and produce it myself, or pass it along to the end user with detailed instructions. How much time would users want to invest? For example, could they be expected to bend their own wire rods? Maybe, but with what tool and what results? Quality could suffer. How much would a wireforming shop charge for a measly 1,200 rods? Ouch! Fallback: Design a jig and pay my son to crank 'em out.

At every turn, I had lots to learn about materials and processes I'd initially assumed were no-brainers. As for steel rule die-cutting, most of my education was over the phone, from guys who occasionally tossed me a technical tidbit while explaining why my job was impossible.

In hindsight, there were other questions I probably should have asked myself. How much of my own time ends up invested in this? Ultimately, is there any payoff? As a fellow DIY-er, though, you'll understand: I must be missing the gene that lets such trivial questions keep me awake nights.

Tim Walker's Digi-Comp I v2.0 kit can be read about and purchased at mindsontoys.com/kits.htm.

Magnetic Switches from Everyday Things

1+2+3 By Cy Tymony

Control many devices from afar with the magnetically sensitive Sneaky Switch.

You will need: **Magnet, paper clip, aluminum foil, tape or foam, cardboard, wire, LED or buzzer, 3-volt watch battery or equivalent**

Optional: **Ring, battery-operated toy, X-10 universal interface and appliance module**

1. Make a magnetic activator.

You'll want a strong magnet to activate devices from at least an inch away. Tiny rare earth magnets can be found in most micro radio-controlled cars, and in the packaging of some hearing aid batteries. Glue a magnet to the face of a ring or a wand, or affix it to some object so that when it's near the switch, or moved away, it will cause the desired effect.

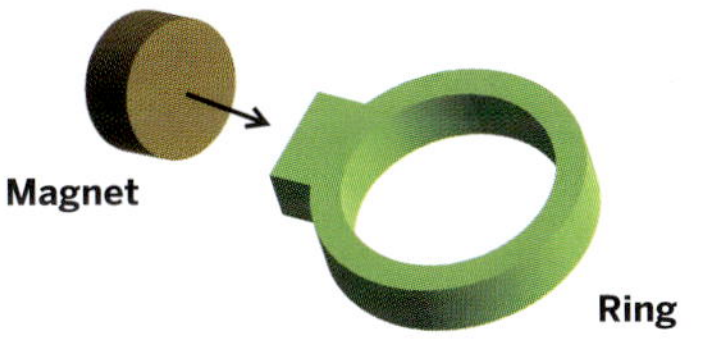

2. Make a Sneaky Switch.

In this magnetic switch, the paper clip lies across a "spring" of rolled tape, one end hovering just above the aluminum foil and the other end taped down. (A small piece of foam can also be used as the spring.) When a magnet passes over the switch, it tugs the clip to touch the foil, completing the circuit. Connect the switch to a 3V watch battery to light an LED, buzzer, or other low-current devices and toys.

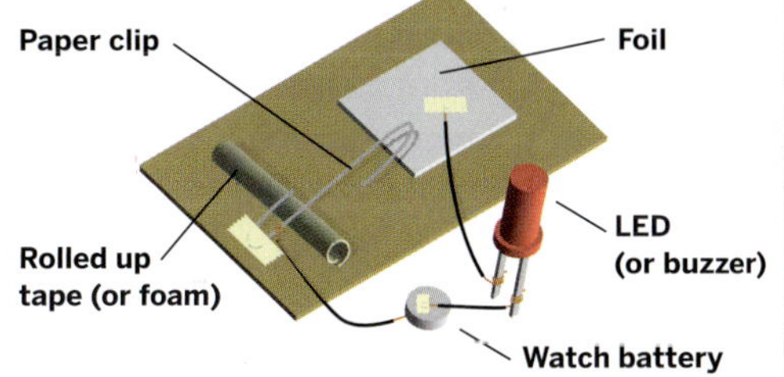

3. Connect switch to a relay.

Your magnetic switch can be attached to a relay to control devices that need higher current. Mount your switch and relay behind the dashboard to secretly activate a cut-off switch, alarm, or other car accessories. Or hook your switch to an X-10 controller and universal interface module to control a variety of appliances. Pretty sneaky!

Bonus: Detect counterfeit money.

Legitimate currency has iron particles in the ink. Fold a bill so half of it stands up vertically — if the top edge moves toward your magnet, it's the real deal. If not, phone the Secret Service!

Illustrations by Mark Frauenfelder

Cy Tymony (sneakyuses.com) is a Los Angeles-based writer and is the author of *Sneakier Uses for Everyday Things*.

Arduino Fever

The tale of a cute, blue microcontroller that fits nicely in the palm of your hand, and the expanding community of developers who love and support it. By Daniel Jolliffe

ARDUINO IS SPREADING RAPIDLY across the globe. But before you reach for the Merck Manual to find the symptoms you're sure to have, check this out. Arduino is actually an open source hardware project that can be programmed to read temperatures, control a motor, and sense touch, and gets its name from an ancient Italian king. And it's fun!

Named after the 11th-century king of Ivrea in northern Italy, the Arduino is both a cute, blue microcontroller platform that fits nicely in the palm of your hand and an expanding community of developers who support it, distributed across two dozen countries, four continents, and counting.

Decidedly 21st-century in its design and construction, the Arduino board is for anyone who wants to build a basic level of intelligence into an object. Once programmed, it can read sensors, make simple decisions, and control myriad devices in the real world.

Using it is a snap: first, hook up a few sensors and output devices to the Arduino, then program it using the free developer's software. Next, debug your code and disconnect the Arduino. Presto — the little blue Arduino becomes a standalone computer.

The original intention of the Arduino project was to see what would happen if community support were substituted for the corporate support that is usually required for electronics development. The first developers — Massimo Banzi, David Cuartielles, David Mellis, and Nicholas Zambetti — ran a series of workshops on assembling the Arduino, giving away the board to stimulate development.

Photography by David Cuartielles (workshop), Ren Wang (robot), and Nicholas Zambetti (chip)

A scant year later, the project has spread far and wide the message that electronic design doesn't have to be a solitary, complex, and painful process, and that it needn't cost much if you have a little help along the way. Says Ren Wang, a student at China's Xiamen University who used the Arduino to power Eye, a walking robot: "Arduino is open and friendly. To make a cool robot was always my dream, and the Arduino made it come true."

Today there's a thriving website with sample code, tutorials, and a forum that serves as the meeting point for Arduino developers. While the original developers still give workshops, the project is increasingly a standalone endeavor, with newcomers taking up the idea that electronics can be a community effort.

Back in Ivrea, a friendly Italian manufacturer, who was courageous enough to support the project from its inception, still provides low-cost Arduinos, in both assembled and kit form. In Europe, the price is €20; Sparkfun (sparkfun.com) is the United States distributor and sells the USB version assembled for $30. And since the project is open source, all the plans, code, and instructions are available online free for those who prefer to roll their own.

Asked what's next, Cuartielles says: "Arduino for kids! We have been asked to evaluate the use of Arduino for technology classes in secondary schools in Madrid, Spain. Can you imagine one million kids a year making experiments in electronics based on this open hardware platform? It would be massive!"

To get in on the massiveness, and to become a contributor yourself, check out arduino.cc.

Arduino assembly workshop (previous page) at Malmo University in Sweden.

"Eye" (this page top) is a robot based on Arduino, designed and built by Ren Wang, a student at Xiamen University in China.

Arduino (bottom) in the palm of a hand.

Daniel Jolliffe is the designer of *One Free Minute*, an anonymous public speech project. He wrote "Throw Your Voice!" in MAKE, Volume 04.

Photography by Howard Cao

Life and Death at Low Temperatures

How to freeze and revive a garden snail. By Charles Platt

Few people realize that the processes of life can restart after long periods in which they are ceased completely. Any backyard biologist can verify this strange fact by performing a couple of experiments that originated in the work of a solitary scientist named Basile Luyet, more than half a century ago.

Luyet was a Swiss Jesuit priest who immigrated to the United States in 1929, did graduate work at Yale, and then set up his own small laboratory in Madison, Wisc., where he toiled in isolation for many years.

He wanted to preserve living creatures in such a way that they could be revived at any time in the future. He knew that this was possible, since some creatures can return from days or even years of lifelessness.

The tardigrade, for instance, is a kind of worm about one millimeter long that is able to withstand temperatures down to almost absolute zero.

Reducing temperature is a logical way to preserve life, because living things are composed of cells, and chemical reactions inside cells occur more slowly when they are deprived of heat.

Even human beings can withstand a moderate amount of cooling. To repair a brain aneurysm, surgeons may reduce normal body temperature by more than 20 degrees Centigrade (36 degrees Fahrenheit), which slows the metabolic rate so that a patient may remain in a state of circulatory arrest for as long as one hour. Brain activity ceases during this period and resumes when the patient's normal body temperature is restored.

THE PROBLEM WITH FREEZING CELLS

However, the freezing point of water forms a barrier below which most mammals cannot go. Ice causes catastrophic damage, although not because it "bursts" cells, as is commonly believed.

When the water inside cells freezes, it forms rigid hydrogen-oxygen bonds that exclude other molecules, such as the sodium and potassium salts that are essential for cellular functions. As the ice accumulates, the normally beneficial impurities are squeezed out and form toxic concentrations that literally poison the cell.

For a graphic demonstration of this phenomenon, place some apple juice in a freezer until more than half of it has turned solid. Now take a sip of the remaining liquid, and you'll find it is the most intense juice you have ever tasted. The water that was mixed with the juice has extracted itself as pure ice, leaving only concentrated juice.

Luyet was well aware of the damage caused when ice forms in cells. He tried to avoid it by freezing tiny tissue samples very rapidly, so that ice molecules would not have time to organize themselves in a crystal structure. As he explained to a visiting journalist, "Living matter can survive freezing, but only if the molecules are not ordered, but solidified where they are ... in disordered or uncrystallized form."

Unfortunately, this technique cannot be used on larger groups of cells, because there is no way to cool them fast enough. In the 1950s, three British scientists who were familiar with Luyet's work solved the dilemma by using glycerol, a form of antifreeze, to replace cellular water and reduce the volume of ice. They described the glycerol as a "cryoprotectant" and named their new field of research "cryobiology," from the Greek word "kryos," meaning "cold." Today, glycerol is routinely used to protect sperm and ova that are preserved in liquid nitrogen. Countless babies have been born from human germ plasm that has endured long periods of lifeless storage with this method.

SAFELY FREEZING AND THAWING AN ORGAN

Human organs represent a bigger challenge because their intricate, delicate structure is easily disrupted by ice crystallization. Luyet managed to revive some fragments of rat hearts, but never recovered a whole organ from a very low temperature. That achievement eluded scientists until 2005, when a team led by cryobiologist Gregory Fahy successfully reimplanted a rabbit kidney after it had been preserved at -130°C. Dr. Fahy spent a large part of his professional life perfecting the cryoprotectant that enabled this feat, which requires a well-equipped laboratory. Still, Fahy points out that backyard cryobiologists can still replicate the very simple demos that Basile Luyet ran more than 50 years ago.

The simplest one doesn't even require glycerol. "I've been told that Luyet would toss two goldfish into some liquid nitrogen," Fahy says. "He would quickly withdraw Goldfish A, which would seem stiff and frozen — but it would resume wriggling after it was placed in warm water."

Luyet would then startle onlookers by removing Goldfish B and snapping it in half. This may not have been a completely fair demonstration, since Goldfish B was exposed to liquid nitrogen a little longer than Goldfish A. It's not clear whether Luyet realized that, to some extent, he had rigged his experiment, but at the very least it remains a unique ... icebreaker?

Liquid nitrogen is available in most urban areas (search online for "liquid gases"), and it is generally inexpensive. It is nontoxic, but must be handled with caution, since its temperature of -196°C can cause serious injury to any exposed human tissue. Always wear heavy gloves and eye protection!

PUTTING A SNAIL INTO SUSPENDED ANIMATION

If you live in a part of the world where snails are abundant, Fahy suggests another simple experiment, which he performed himself as a high school student. All you need is a domestic freezer and a thin-walled vial, such as the kind used for prescription medications. First, catch your snail. This is the easy part. Now place the snail in the container, top it off with water, and put it in your freezer. Wait just long enough for the water to turn to ice (an hour should be sufficient). Quickly remove the container and run it under warm water. As the ice melts, transfer your snail to a smooth surface such as a dinner plate, and continue the warm-water treatment.

When the snail recovers its senses, it should start crawling away at a fast pace (relatively speaking). If your snail turns yellow and breaks in half, you left it in the freezer too long. Even snails become injured when they are 100% frozen without protection from ice damage.

Those who want to take cryobiology more seriously can obtain some glycerol from a chemical supply company. (Glycerol is nonhazardous as long as you refrain from drinking it.) If you have access to a cheap microscope, you can replicate British research from the 1950s by soaking sperm or blood cells in various concentrations of glycerol. Subject your samples to liquid nitrogen, then rewarm them and see if the cells resume their activity.

If you're feeling even more ambitious, you can perform the same procedure with nematode worms, which are as small as tardigrades but more readily available. In fact, your backyard may be crawling with them. You can also buy them by mail order from ecologically enlightened companies that sell them as a natural pest control, since the microscopic worms eat fly larvae. Reviving a nematode from a period in liquid nitrogen is a challenge, but it can be done.

Basile Luyet never achieved his ambition to stop and start life processes on a large scale, but his basic studies had serious long-term implications.

Clearly, a surgical patient who reawakens after zero brain activity at a low temperature has not been dead in the usual meaning of the term. By the same logic, a frozen blood cell, or a cryopreserved nematode, is not dead either. If we can revive a rabbit kidney, or (eventually) a whole animal, by preserving it in such a way that its cells remain viable, we may begin to challenge the conventional concept of death to the point where it becomes virtually meaningless.

FURTHER READING

societyforcryobiology.org
21cm.com
alcor.org

Charles Platt has been a senior writer for *Wired* and has written science fiction novels such as *The Silicon Man*.

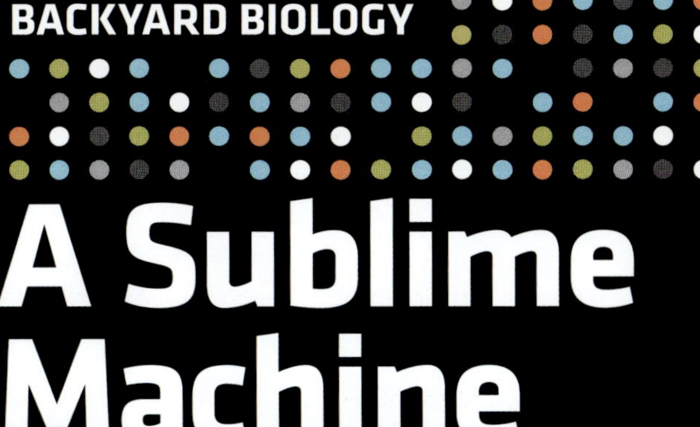

BACKYARD BIOLOGY

A Sublime Machine

Mike Wilder makes Lego robots for time-lapse 3D videos of carnivorous plants. By Mike Kuniavsky

Mike Wilder recalls: "When I was 7 or 8, we went to EPCOT and saw an amazing 3D film that was produced by Kodak. In one of the scenes, a character threw a gold ring at the audience. Everyone in that audience reached out simultaneously to grab the ring. That just blew my mind."

Twenty years later Wilder created his own 3D science movie. Working alone in the closet of his one-room basement apartment with less than $500 worth of materials, he made the first 3D time-lapse movie of tiny carnivorous plants, using robots made of Lego.

It all started with the plants. Wilder had never grown anything before he bought his first American Pitcher Plant for $5 at a store in San Francisco in 1995. Ten years later, his collection includes 200 different varieties of carnivorous plants from all over the world.

However, collecting plants was not enough for him. Wilder has been breeding the Mexican *Pinguicula* plant to produce new kinds of flowers. Wearing a jeweler's magnifying visor and holding tiny tweezers, he touches the miniature sexual organs of one breed of *Pinguicula* to those of another (*see Hacking Your Plants, page 72*). If fertilization takes place, he collects the seedpods and sprinkles the seeds into special soil he makes for each plant variety. If he's lucky, the plants will flower three to five years later.

Wilder is clearly a patient man, but watching carnivorous plants grow is not very different from watching grass grow. That's when the image from EPCOT came back. He decided, "I want to see this in 3D. I want these plants to be spinning while they're growing." But there was a problem: no one had ever done 3D macro time-lapse photography with rotation (as he, also an avid collector of time-lapse plant movies, knew). Why? "Because it's hard," says Wilder. Standard lenses are too big, and he didn't have the budget for specialized gear.

After pondering the dilemma for a year, he arrived at the solution when he saw the work of another Portland 3D artist, Vladimir, who makes her own View-Master reels using a single camera. "That gives me the loophole I need. Since my particular subjects move so slowly, I can use my digital still camera and build a robot that will have the camera in one position, take a picture, and then slide it over

Photography by Mike Kuniavsky

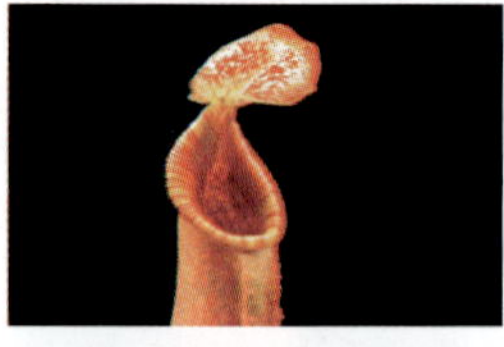

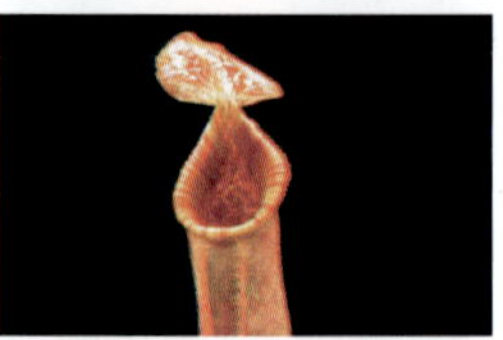

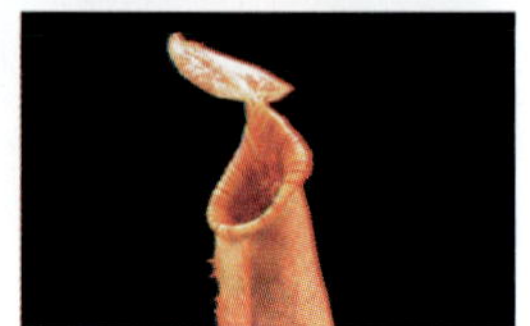

Wilder's 3D camera. Taking a cue from the infinite resourcefulness of a natural world that can produce something as wondrous as a carnivorous plant, he says his goal was to "make a sublime machine."

about 5mm [to take another picture]."

The next challenge: Wilder had never built a robot. He checked out every book on robotics and Lego from the public library and spent his evenings reading. There were many more challenges to overcome: "Can I get Lego to do 1/10 of a degree rotation? Can I get a Lego car to drive 1/10 of a millimeter? Will the Lego microcomputer tolerate being on 24 hours a day for two weeks? For a year? Will my camera?" Lighting was also a problem: plants (and the insects they may eat) need light to grow, but it may not necessarily be good light for cinema.

His consultations with experts were not encouraging. "I had a guy who had a Ph.D. in physics tell me, 'You're insane. Even if you had a real robot, it would be really hard to pull this off. There's no room for error. If the two camera positions are A and B, it always has to be A and B because you're trying to create the illusion that a 3D camera is on a tripod. It needs to be A and B *every single time*. You're an idiot if you think you can do that with Lego.'"

"I had a guy who had a Ph.D. in physics tell me, 'You're an idiot if you think you can do that with Lego.'"

Wilder was determined to prove the naysayers wrong. "I don't care if this is an impossible movie. I don't care if the physicist says it can't be done. IMAX has not dared to do it, or the BBC. *I'll* do it. And I'll do it with Lego because that will be beautiful." He succeeded. The robot's name is Jasper.

Controlled by a single Lego Mindstorms RCX brick, Jasper consists of three parts: the *juxtaposition module* moves his Nikon Coolpix 4500 camera back and forth, the *rotation module* rotates a flower pot, and the *cable release module* pushes the trigger on a camera remote to take the actual pictures. Wilder composited the alternating left- and right-eye images into a single movie using a piece of Japanese freeware that adjusts the colors to use classic red-and-blue 3D glasses.

After 18 months of filming, scripting, animating, and editing, Wilder produced a 20-minute DVD called *The Carnivorous Syndrome in 3D*. Although subtitled "Our curious universe, science adventure no. 5," it's not a parody, but an earnest attempt at updating the genre. The soundtrack to the piece consists of ambient electronica, and the graphics have the cool detachment of a minimal techno CD.

The Carnivorous Syndrome in 3D is available as a DVD directly from Wilder's website, 3dsyndrome.com.

Mike Kuniavsky is a San Francisco- and Portland-based ubiquitous computing and user-experience consultant and writer who blogs at orangecone.com.

BACKYARD BIOLOGY

Kitchen Counter DNA Lab

Extract, purify, and experiment with the blueprint of life.

By Dr. Shawn

Photography by Howard Cao

DNA is perhaps the most extraordinary structure in all creation. Its famous double helix is the longest molecule known and regulates the life processes in every cell on Earth. Even more, the code that DNA carries is the actual blueprint of life itself. The human recipe, for example, consists of roughly 3 billion molecular bits of information laid out in a precise sequence. Perhaps most amazingly, this miraculous winding staircase directly links every creature on Earth to our ancient and common past — far back to when evolution first began shaping the biological forms that would ultimately populate the world we know today. By examining the differences between the DNA in our bodies and that in other organisms, we can tell when our species diverged from chimps, apes, and even primordial fish.

The properties of this massive molecule are so mysterious and wondrous that most folks assume only the enlightened priesthood of laboratory biologists can extract and study it. Not so. In fact, anyone can extract, purify, and experiment with DNA at home.

When released from a cell, DNA typically breaks up into filaments. In solution, these strands have a slight negative electric charge, which makes for some very useful chemistry. For example, the more negative sections of one DNA strand will tend to attract the more positive regions of another. This causes DNA molecules to clump together and fall out of solution. However, if salt is added, its positive ions are attracted to the DNA's negative charges, effectively neutralizing them. This stops the fragments from adhering and keeps them floating in solution.

So, by controlling the salt concentration, anyone can make DNA fragments either disperse or clump together. And therein lies the critical secret of separating DNA from cells and manipulating it at home.

ISOLATING THE DNA

Extraction

Here's how it works. First, you'll need a salty solution, called a buffer, into which DNA can dissolve. Next, you'll need to break open a bunch of cells and let their molecular "guts" seep out into your buffer. Then, you'll want to add a special enzyme that will destroy unwanted molecules, such as proteins, which would otherwise contaminate your results. Finally, you'll have to reduce the salt concentration enough to cause the DNA molecules to clump together and fall out of solution.

Step One: Build the Buffer

First, you'll need to whip up your buffer (see recipes on next page). Pour 120ml (about 4 oz.) of distilled or bottled water into a clean glass container. Add the table salt and baking soda, and stir vigorously. After they have dissolved completely, stir in the detergent. Shampoos and liquid laundry detergents that contain sodium lauryl sulfate (check the label) work well.

Next, add the tenderizer by wetting a toothpick, inserting it into the meat tenderizer, and transferring it to the buffer. Meat tenderizer contains an enzyme called papain that breaks up proteins so they won't come out with the DNA. Pineapple juice and contact lens cleaning solution also contain protein-busting enzymes, so, alternatively, you can add a drop of one of these two liquids.

Lastly, because DNA degrades fast (sometimes in a matter of minutes), you'll want to slow the pace of destruction by chilling the buffer in a bath of crushed ice. If the buffer becomes cloudy, you've chilled it too much. In that case, warm it just enough to clear it.

Step Two: Get the DNA

For a source of DNA, try the pantry. You can get great results with raw onions, garlic, bananas, or tomatoes. But it's your experiment; choose your own personal favorite fruit, veggie, meat (fresh or frozen), or fungus.

Once you've secured your DNA source, you'll need to process its cells to extract their organic molecules. First, use a knife to dice the material into small pieces. Put the material into a blender and pour in just enough distilled or bottled water

For the buffer: **Distilled or bottled water (glass 1)** 120ml (about 4 oz), **salt** 1.5 grams (¼ tsp), **baking soda** 5 grams (1 tsp), **liquid laundry detergent, dish detergent, or shampoo (glass 2)** not soap — look for sodium lauryl sulfate on the label, 5ml (1 tsp), **crushed ice** to chill the buffer, **meat tenderizer, pineapple juice, or contact lens cleaning solution** just a dollop.

For a source of DNA: **Anything with living cells or cells preserved by freezing such as** fruits, vegetables, legumes, fungi, meat from the butcher shop (a frozen cow tongue works great!), bone marrow from soup bones, etc.

To extract the DNA: **Isopropyl (rubbing) alcohol (glass 3)** with no additives and as concentrated as possible. Chill the bottle in the freezer before you begin.

Sundries: **A drinking glass** to mix the buffer, **small narrow glass container** (preferably with straight walls; a test tube is ideal, but a shot glass will do) to extract the DNA, **narrow drinking straw** to add the alcohol, a **graduated test tube** (or a plain one and a ruler with a centimeter scale) to measure the DNA, **glass swizzle stick** to remove the DNA.

to cover the chunks. Then break up (or *lyse*, as biologists say) the cells by pulsing the blades in short bursts until you've blended the material into a slushy mass. This will rip open some of the cells directly and expose many more cell walls and nuclei to the detergent's attack.

Finally, you need to leach out the organic molecules. Place 5ml (1 tsp) of the minced mush into a clean container. Mix in 10ml (2 tsp) of your chilled buffer. Swirl *gently* for 2 minutes, and the guts of the shattered cells will separate into the buffer intact. If you stir too vigorously, you'll break up some of the DNA.

Step Three: Dump the Gunk

Next, you'll want to separate the solid gunk from the molecule-laden soup. This is best done with a centrifuge. If you don't have access to one, you can always build one yourself from an old kitchen blender. (To learn how, check out "A Kitchen Centrifuge" on my Tail-Kicking Downloads page scifair.org). If you choose this route, spin at low speed for 2 minutes.

If you don't have a centrifuge lying about and don't want to build one, there are simpler options, such as the toe of an old nylon stocking. Just cut 6 inches off the foot, drop the toe into a clean drinking glass or jar, stretch the fabric across the opening, and pour your molecular broth through. The stretch fabric will cling to the glass and the fine mesh makes a wonderful filter.

Step Four: Extract the DNA

When you've extracted the liquid from the gunk, carefully pour at least 5ml (1 tsp) of the fluid into a narrow vessel, such as a clean shot glass, clear plastic vial, or test tube. (If you're using a vessel larger than a test tube, you'll need more fluid. Use enough to fill the container at least one-quarter full.) You are now ready to coax the DNA molecules to stick together and fall out of the solution.

Remember, the DNA is only suspended in the buffer because salt ions prevent these giant negatively charged molecules from sticking together. Now, you're ready to reduce the salt concentration enough to let the DNA molecules clump together and fall out of solution.

Remove your chilled alcohol from the freezer. Along the inside of the container, you'll need to carefully pour about the same amount of alcohol as you have buffer, so that the alcohol gently settles on *top* of your DNA-laden buffer. To do this, dip a narrow drinking straw into the alcohol bottle and then block off the top of the straw with your finger to capture some alcohol. Remove the straw, tilt the glass, and touch the tip to the inside of the glass. Then, simply let the alcohol flow down along the side. Because the alcohol is less dense than the buffer, it will float on top. If you have a very steady hand, you can also do this step by gently decanting the alcohol into the container by pouring the solution down along a pencil into the container.

Where the 2 liquids meet, a gelatinous sludge will suddenly appear. That sludge is DNA!

At this point, you should see 3 distinct layers: the alcohol on the top, the DNA sludge directly below that, and the buffer on the bottom. The DNA should appear as stringy filaments that stick together. If, instead, it appears as chunky pieces of floating debris, something happened to break up the molecules. You'll still be able to measure its volume, but you may not be able to remove it for study.

Buffer Banter

In the lab, scientists often use the detergent sodium dodecyl sulfate, or SDS, to extract DNA from cells. SDS is also common in shampoos and household detergents, where it goes by the name sodium lauryl sulfate.

Scientists also use table salt – pure sodium chloride – without additives. Morton ("When it rains, it pours") adds calcium silicate to its brand of salt to prevent caking in high humidity. But too many calcium (or magnesium) ions can react to lace your buffer with a white "soap scum," especially if you use a soap rather than a detergent. You can use a liquid soap, but you'll need a salt with no calcium or magnesium compound added (read the label). Water softener salts (both sodium chloride and potassium chloride) work well. Otherwise, use detergent and table salt.

Scientists also use distilled water but bottled water will work just fine. Just don't use tap water because it's loaded with undesirable ions and (often) worse, chlorine, which destroys DNA on contact.

Dyeing DNA

Some dyes bind directly to DNA. Adding a drop or two to the solution will stain the transparent sludge so you can easily see your entire harvest. The safest of these for home use is methylene blue. It is non-toxic, and since it is used to treat certain illnesses in fish, you can find it at many well-stocked aquarium supply stores. It usually comes as a 2.3% solution. The proper DNA stain is about a 1% solution, so you'll want to dilute it by adding an equal amount of bottled or distilled water.

TAKING IT FURTHER

DNA Experimenting

WELCOME TO THE WORLD OF MOLECULAR BIOLOGY

There are two types of experiments that are particularly easy to do, and I recommend that even the most adventurous experimenters start with one of these: discovering how much DNA can be extracted from different organisms under different circumstances, and exploring the conditions that cause DNA to degrade.

Measuring the amount of DNA you've extracted from a sample couldn't be simpler. First, measure the inner diameter of your straight-walled container holding the DNA. Once you know that number, just measure the thickness of the sludge. With that information, it's easy to calculate the volume of DNA you've produced: the equation is $V = \Pi D^2 T/4$, where D is the inner diameter of the vessel and T is the thickness of the layer of DNA sludge. Next, divide the volume of extracted DNA by either the volume or the mass of the material it came from. The simplest way to do that is to accurately measure how much mush you put into your buffer, and then process all of the buffer to extract every scrap of the DNA that leached into it. If you have an accurate scale, weigh your sample before processing it. If not, just measure the volume of the material before you blend it.

Example: Suppose you processed 5g of onion into 10ml of buffer and extracted 1ml of DNA. How much DNA did you get from each gram of onion? Easy! Just divide what you got by what you started with: 1ml DNA / 5g onion = 0.2ml/g.

You can also run experiments with the DNA itself. Usually, the first step is to remove the DNA sludge. It takes a little practice, but you can do it using a clean glass and a swizzle stick. Gently insert the stick through the layer of alcohol and swirl it very slowly in the same direction, with the tip suspended just below the top of the buffer solution. Longer pieces of DNA will spool into the glass, leaving smaller molecules behind.

After a minute of swirling, slowly pull the stirrer up through the alcohol. This will make the DNA adhere to the end of the stick, where it will appear as a transparent, viscous, "snotlike" clump clinging to the tip.

If you now resuspend the molecules in a fresh batch of buffer, you can expose them to chemicals, sunlight, temperatures, or anything else that might break up the DNA. Make a new batch exactly as before and chill it, but don't bother adding the detergent. Submerge the swizzle stick and gently agitate for several minutes as the DNA dissolves into the buffer. Then divide the buffer equally into 2 clean glass containers. Expose one — your test sample — to whatever agent you wish to test. Leave the other — your control— alone. Then process both as quickly as possible and compare the amount of DNA you can extract from each buffer. The difference between the volumes from your test and control is a measure of how much damage your agent does to DNA.

If this sounds too easy, keep in mind that DNA is fragile stuff and it can be affected by lots of subtle things that might escape your notice. Getting consistent results takes practice, so make sure you vary the exposure and that your plotted data shows a regular behavior before drawing any conclusions.

COLD STORAGE

It's actually easy to store your DNA for later use. Just place the "snot bulb," swizzle stick and all, in a container filled with ice-cold isopropyl alcohol and put it in the freezer. Your DNA will keep almost forever.

Dr. Shawn (Shawn Carlson, Ph.D.) is a MacArthur Fellow and the founder and executive director of the Society for Amateur Scientists. To learn more about the society, visit sas.org.

Home Molecular Genetics

Extract, fingerprint, and replicate your own DNA.

By the UBC Advanced Molecular Biology Lab: Jon Nakane, Keddie Brown, Peter Danielson, Joanne Fox, Yas Shirazu, Donna Lee, Esther Abd-Elmessih, and David Ng

Working directly with DNA isn't only for the labs of CSI, agribusiness, and headline-grabbing research institutions. It's basic chemistry, but it uses the molecules of life. You can even do it at home. This article explains how you can isolate and "fingerprint" some of your own DNA (which is easy), and replicate enough of it to perform more accurate and detailed fingerprinting (which is a bit more difficult). You can view these experiments as an extension of high school education, a low-cost contribution to science infrastructure in developing countries, or perhaps even an exercise in bioethics.

PROJECT 1:

Extract and Characterize DNA

First, we'll extract some DNA from our cheek cells, using a process similar to the one described in Shawn Carlson's article on page 59 (*Kitchen Counter DNA Lab*), which goes into greater detail. Here's how to do it.

You can find the full materials list for this extraction method online at makezine.com/07/fingerprinting.

Rinsing buffer: 1.5g (approximately ¼ tsp) table salt, 5g (about 1 tsp) of baking soda, and ½ cup bottled water.

Running buffer: .05g (a pinch) of table salt, 2g (½ tsp) of baking soda, and up to 1L bottled water. Use a pet-store aquarium kit to make sure this buffer measures pH 7.5; add water to lower the pH, or add baking soda to raise it.

Loading buffer: 1.25ml (¼ tsp) glycerol/glycerine (available at most pharmacies) and several drops of red food coloring.

Mix the Buffer Solutions

Mix 3 buffer solutions according to the recipes listed above.

Extract the DNA

Swirl 5ml (1 tsp) of rinsing buffer around in your mouth for about 30 seconds. Swirl gently to prevent the sample from becoming frothy. The more cells you can rinse out, the better — so don't do this after brushing your teeth.

Spit the buffer into a paper cup and pour it into a test tube. Try not to spit out more (from added saliva) than the original volume of rinsing buffer. Squirt a bit of liquid soap (about ¼ tsp) into your sample and mix well, but gently, with a wooden toothpick. Slowly add 5ml (1 tsp) of cold rubbing alcohol to the sample, pouring it down the side of the test tube at an angle so you form 2 undisturbed layers of liquid.

After 10 minutes, the DNA should appear as a whitish, snot-like substance floating between the rinse solution layer and the alcohol. Results can

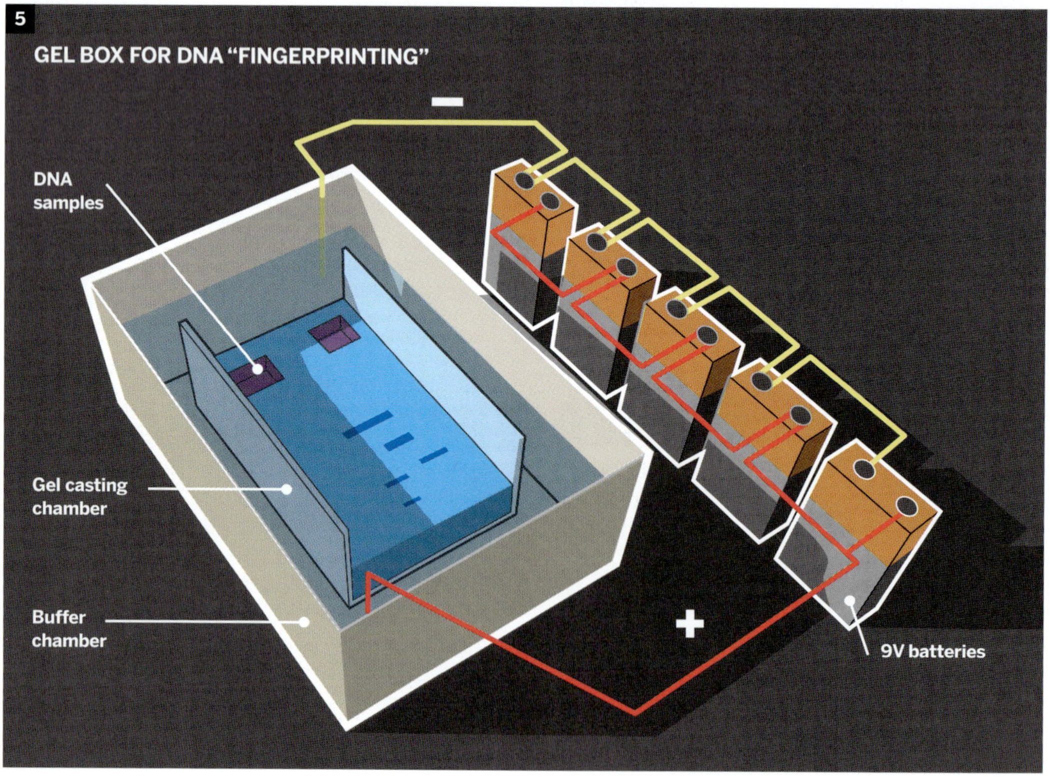

1

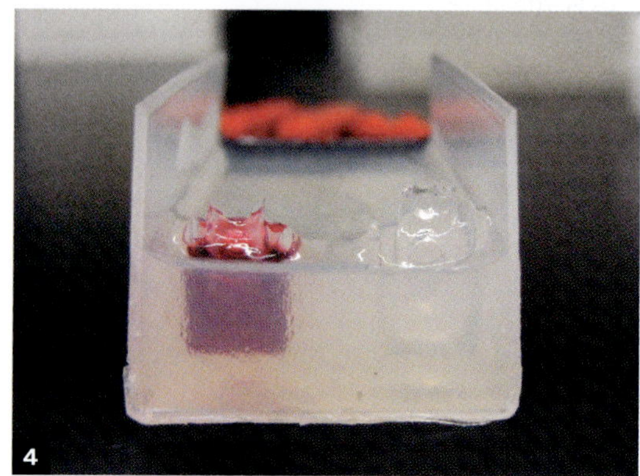

5

GEL BOX FOR DNA "FINGERPRINTING"

–

DNA samples

Gel casting chamber

Buffer chamber

+

9V batteries

Illustrations by Kirk von Rohr / Photography by the UBC Advanced Molecular Biology Lab rinse solution layer and the alcohol. Results can depend on the soap you use and other factors, so you may have to experiment.

Make the Gel Box

The gel electrophoresis process lets you separate and visualize different DNA molecules based on their sizes. Agar gel — a dense, microscopic network of sugar — is difficult for larger DNA molecules to travel through, but because DNA is negatively charged, you can draw it through the gel by applying an electric field. A gel box uses this principle to act as a DNA racetrack. The shorter the strand of DNA is, the faster it runs.

The box consists of 2 nested containers. An inner "casting chamber" contains the gel itself and sits in the running buffer inside the outer "buffer chamber." Two opposite ends of the casting chamber are open to give the gel direct contact with the electrolytic buffer. Two corresponding electrodes in the buffer chamber electrify the liquid across the 2 ends.

We built our box in 2 ways: using half of a plastic travel soap dish inside of a larger rectangular Tupperware container, and using the lid of a plastic slide box inside a box made of Lego blocks and lined with Glad Press 'n Seal wrap to prevent leakage. In both cases, we cut away the narrow sides of the casting chamber, creating a U-shaped channel. For the electrodes, we ran wire along both ends of the buffer chamber and connected each to a screw as a contact.

Prepare the Gel

The gel needs 2 wells at one end, in which to insert the DNA; we'll make these by casting the gel around a 2-pronged Lego "comb" sticking down into the casting chamber. Make the comb as shown (Figure 2), then seal the open ends of the chamber with masking tape (Figure 1).

On a hot plate, heat 8g (about 2 tsp) of agar and 125ml (½ cup) of running buffer in a small pot. Stir gently until the agar is fully dissolved and the solution is clear, which may take up to 30 minutes. Pour the liquid into your gel-casting chamber to a thickness between 0.5cm to 1cm. Hang the comb into the gel at one end and let the gel cool (Figure 3). After it sets, remove the comb and masking tape.

Prepare the DNA

Use a wooden toothpick to spool up the DNA from the extraction step. Mix 7 parts rubbing alcohol to 3 parts water in a small container, and dip the DNA "snot" into this solution for a few seconds. Air-dry the glob on the toothpick for 10 minutes, then scrape it into 75ml (3 droplets) of running buffer and allow it to dissolve overnight at room temperature. Add 1 drop of loading buffer, and your sample is ready to run.

Run the Gel

Load as much of your DNA sample as possible into the wells of the gel (Figure 4). Put the gel-casting chamber into the buffer chamber and cover it with running buffer until it's fully immersed, with 0.5cm of fluid above the gel. Connect the electrodes to 5-7 9V batteries, wired in series (Figure 5). Since DNA is negatively charged, be sure to position the positive electrode at the end opposite the wells.

Let the race begin! The batteries should supply enough power for a 3-hour run, but the exact timing depends on the dimensions of the gel, buffer chamber, etc. The smaller the chambers, the faster the run (due to less resistance). Do not stick your finger into the fluid during the run. You can check if the circuit is complete by looking for bubbles forming on the positive electrode.

Stain the Gel

Make a 0.02% solution of Methylene Blue in distilled water; some aquarium supply and pet stores carry Methylene Blue in 2.3% solution, so you'll need to dilute this at about 12:1. Immerse your gel chamber in the solution overnight at room temperature. If all goes well, you should be able to see at least one faint band of DNA from your cheek cells.

This DNA "fingerprint" process is not good enough to match a sample with the person who submitted it, but it could probably be used to determine whether the DNA came from a human rather than, say, a banana. In fact, you can make a fingerprint using banana or bean DNA (*see page 59*) and compare the differences with the cheek-cell sample.

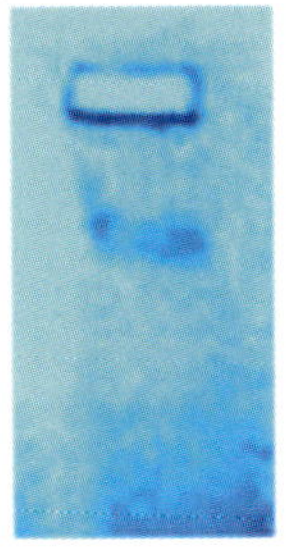

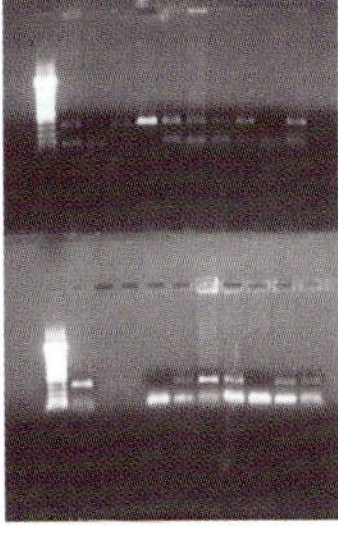

The DNA fingerprint on the left was created using the process outlined in this article. The DNA fingerprint on the right was made using laboratory quality PCR (polymerase chain reaction). In the next section of this article, we'll show you how to make your own PCR system.

PROJECT 2:

Build a Thermal Cycler and Run PCR

Real DNA "fingerprinting" is usually done using a procedure called *polymerase chain reaction*, or PCR. This process replicates DNA, making a much larger sample that produces more detail in electrophoresis and is therefore easier to match. To perform PCR, you need some specialized chemicals and equipment.

The chemicals are small, predesigned pieces of DNA known as primers, plus a heat-stable DNA polymerase reagent such as Taq. The primers combine into new copies of the sample DNA strand, and the polymerase enzyme catalyzes this assembly process. Both of these materials are readily available from biotech supply companies such as Takara (takaramirusbio.com) for less than $100, which buys enough for about 100 reactions.

For the hardware, you need some small plastic tubes and a thermal cycler, which applies programmable temperature changes to the tubes. Commercial thermal cyclers for laboratories range from $2,500 to more than $7,000, but you can make your own MacGyver version using a Handy Board microcontroller (handyboard.com, around $225) and about 50 dollars' worth of additional components. Here's a metalevel description of the different pieces and how to put them together. (You can find the schematic and full parts list online at makezine.com/07/fingerprinting.)

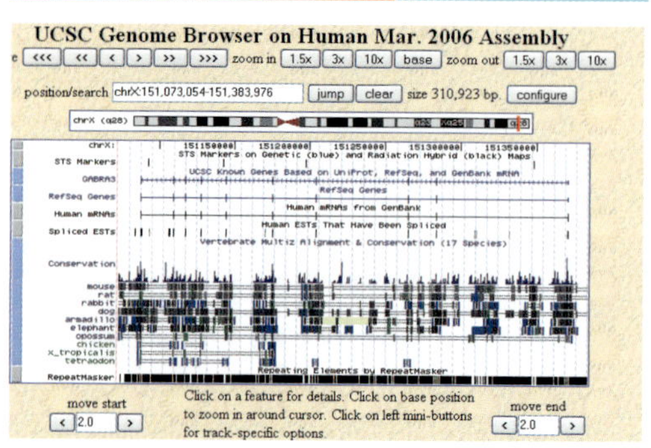

THE GENOME BROWSER

Our genome represents all of the DNA in our cells' nuclei. This DNA is the "genetic blueprint" that determines how we're put together on a molecular level, what we look like, and how healthy we are. It contains over 3 billion letters, called *nucleotides*, which the Human Genome Project has mapped using DNA sequencing technologies built from the same basic principles outlined in the projects presented here. Now that we have the sequences, the next step is figuring out what they do, which parts of the sequence aren't "junk" and actually produce proteins, and what these proteins' functions are in the body.

Anyone can read this blueprint and browse the latest discoveries online using the Genome Browser at the UCSC Genome Bioinformatics Site (genome.ucsc.edu/cgi-bin/hgGateway). This breakthrough tool is like a Google Maps for genomes, and it's being updated continuously as researchers decipher different parts of the genome.

You can use the Genome Browser to search the entire genome sequence and navigate around any part of it. You can see the detailed features of any particular location by searching for an address; instead of a street address, you enter the numerical position of the nucleotide in the entire sequence. Researchers routinely use the Genome Browser when they need raw data from the human blueprint.

How It Works

The component that performs the thermal heavy lifting for our thermal cycler is a Peltier device, aka thermoelectric cooler. This is a flat, solid-state device that "pumps" heat from one plate to the other when you apply a DC current. Inside, current zigzags through alternating sections of P-type and N-type semiconductor material sandwiched between the two plates. Heat is drawn away along the plate, where the current jumps from P to N; along the other plate, where current runs from N to P, heat is given off. If you reverse the direction of the current, you reverse the heat flow. Peltier devices are used to cool microprocessors and photoelectric devices. By themselves (without a power supply and controller), you can get them for less than $15 from surplus companies; check peltier-info.com/surplus.html. We used a 1.5"×2" device rated at 5V and 8 amps (Marlow Industries item #SP2083). For the device's power supply, we used a conventional benchtop power supply, but the power supply from an old computer would also work.

How to make a thermal cycler for DNA replication:

1. Instead of paying $5,000 for a commercial thermal cycler (which you'll need to replicate DNA samples), you can make the one shown here for less than $300.

2. This Peltier device pumps heat from one plate to another when a current is applied.

3. Here, the Peltier device has been outfitted with 2 aluminum blocks. The top block has holes to hold the reaction tubes. The bottom block is a heat sink.

4. The top block of the thermal cycler contains holes for a tube where the reaction takes place, and a dummy tube that contains a temperature sensor. The data from the sensor provides feedback to the microcontroller, which controls the Peltier device.

4

THERMAL CYCLER FOR PCR

Dummy tube

Reaction tube

Temperature sensor

Mineral oil

Thermoelectric element

Heat sink

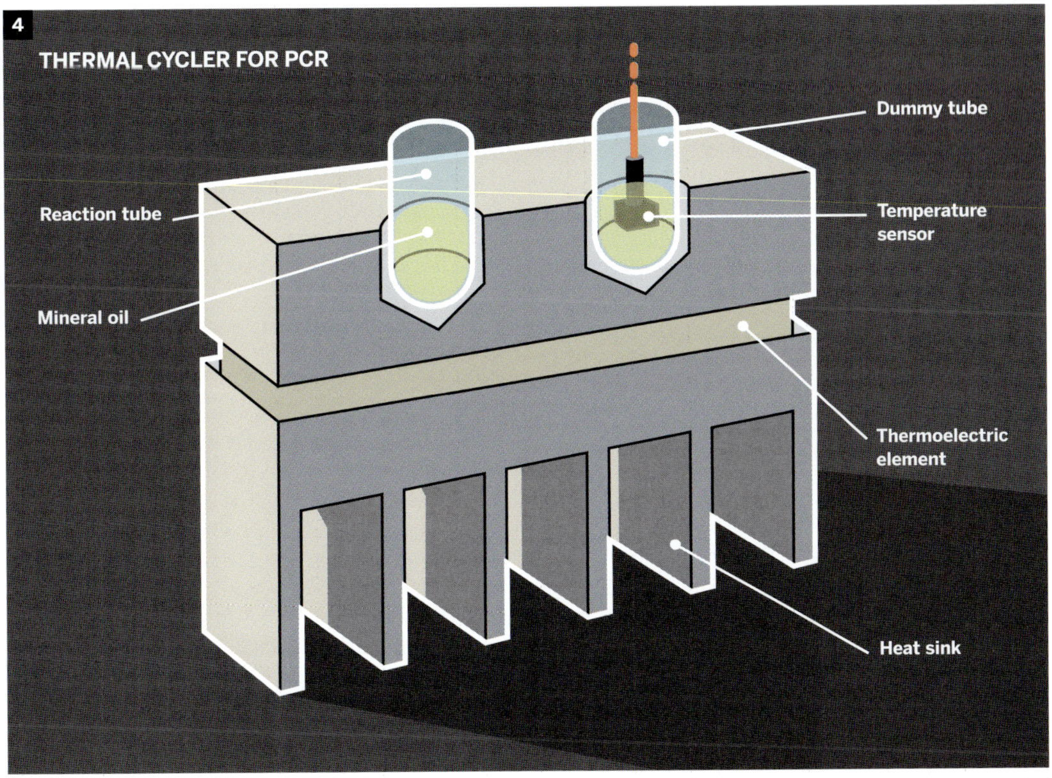

The Peltier device is sandwiched between 2 chunks of aluminum. A thermal block on top holds and transfers heat into the tubes. To ensure good thermal contact between the block and the tubes, fill the holes with mineral oil. A heat sink underneath absorbs heat and radiates it into the surrounding air. All 3 components need to be tightly screwed together with nylon screws, using thermal paste for optimal and uniform heat transfer. We drilled 2 holes in the top of the thermal block, one for holding our sample and the other for the temperature sensor.

A temperature sensor inside one "dummy" tube monitors the temperature inside the other tube. We used a National Semiconductor LM335Z, which produces a small voltage that changes proportionally to its current temperature. Because the voltage changes are slight, our circuit amplifies the output with a Burr-Brown OPA 4241PA op-amp chip. Three potentiometers in the sensor circuit let you adjust the circuit's output signal so that it ranges between an even 0V and 5V for feeding back to the microcontroller. The Peltier device operates within an H-bridge subcircuit. This configuration of 4 switches — arranged like an "H" — lets you run current in either direction through a device.

H-bridges are often used with DC motors, but we're using it to heat or cool our sample. The Handy Board has on-board H-bridges, but they can only handle 600mA. To switch our 8-amp Peltier current-hog, we take the motor output signal from the Handy Board and pass it to our own external H-bridge circuit constructed out of a comparator, 2 bipolar junction transistors, and 2 high-current MOSFET transistors.

The microcontroller follows the thermal-cycling program and monitors input from the sensor circuit in order to provide the proper amount and direction of current to the Peltier device. Many standalone boards could do the job, but we chose MIT's Handy Board (handyboard.com), which some of us had used previously for an Engineering Physics course. The Handy Board has analog and digital inputs, an LCD, infrared communications, and, most importantly, supports fast application development via Interactive C.

The code for the microcontroller, which we wrote in Interactive C, follows the prescribed thermal-cycling recipe and maintains the proper temperatures. Here's pseudocode for the main program loop:

```
For (number of complete temperature cycles)
{
  For (each target temperature in the cycle)
  {
  Get reading from temperature sensor
  While (temperature sensor is not equal to the target
temperature)
            Apply current to cool or heat the sample to reach
target temp
  Start timer
  While (target temperature reached and timer running)
            Apply current to cool or heat sample to maintain
target temp
  }
}
```

On the following page, we'll look at how our PCR system replicates DNA. You can rerun the gel electrophoresis experiment using the replicated DNA. Does the fingerprint look different with DNA made from PCR?

ETHICS: GENETIC TESTS AND TOYS

Genetic tests pose significant ethical questions. The MacGyver DNA extraction, electrophoresis, and replication techniques described in this article permit real genetic experiments, but they're still very crude compared to clinical genetic tests.

Unless you try really hard with these techniques (remember: don't eat the chemicals and watch out for the electric current), you won't step into the ethical danger zone that professional genetic testers and takers face. Nevertheless genetic testing raises three noteworthy ethics issues:

1. Our MacGyver experiments might distinguish you from a banana. More accurate professional tests can unhinge your future or your family by revealing life-changing information.

2. Even accurate genetic tests produce only probabilistic information. Most of us don't deal with probability terribly well — witness Las Vegas and "zero-tolerance" policing. Since so much is at stake with a genetic test, you should consult a physician or genetic counselor for interpretation and advice.

3. Genetic information can be used strategically against you. Employers, insurers, and others can harm you if they discover your results (risk status) or simply the fact that you had a genetic test. This is a further reason to consult a medical professional, whose code of ethics protects your privacy.

These risks do not undercut the benefits of some genetic tests. Bryn Williams-Jones' Genetics and Ethics page at genethics.ca is a good place to start further study. ***—Peter Danielson***

Running PCR

You perform polymerase chain reaction (PCR) by mixing your sample with polymerase reagent and primers in a small test tube, then running it repeatedly through a sequence of temperature changes in your thermal cycler. Each temperature change facilitates a different step in the replication process, and the entire sequence doubles the volume of DNA in the tube that's identical to the original sample. Repeat the process 30 or 40 times, and you'll end up with a billion or trillion-fold increase in your DNA, as long as your primer molecules hold out.

Gel electrophoresis, as outlined in the first part of this article, determines the DNA's "fingerprint," and the large amounts of DNA from PCR make these fingerprints easy to match. Run identical gel electrophoresis on generously replicated samples, and if they produce a similar set of lines on the gel, then the samples match — although it's still possible (if unlikely) that the matching lines come from different DNA strands with the same lengths and weights.

1. Open the DNA.
(Thermal cycle: about 95°C for 1 minute)
DNA exists as a tight helical coil, and applying a high temperature unravels it.

2. Primers bind.
(Thermal cycle: about 60°C for 1 minute)
With the DNA open, shorter pieces of DNA called primers — designed to interact with a region you wish to copy — can now get in and bind to each half. The second temperature encourages this binding reaction.

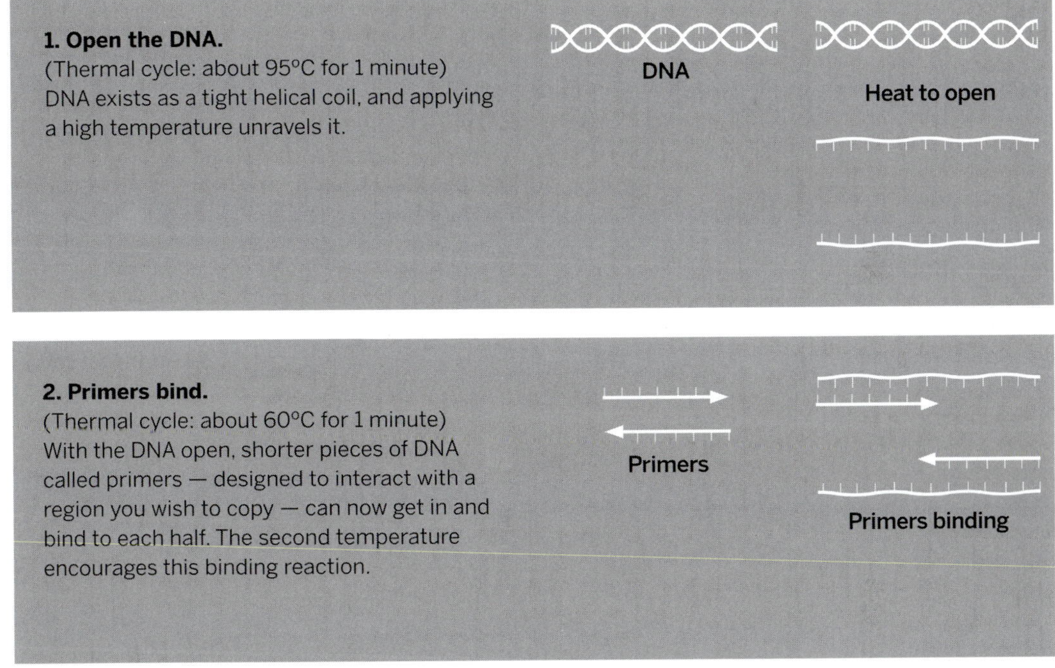

3. Polymerase extends replication.
(Thermal cycle: about 70°C for 1-3 minutes)
With the primers bound and providing a foothold, the heat-stable DNA polymerase runs along the half-DNA original and extends replication of the other half. This is the actual "replication" part, and the new temperature optimizes enzyme function. At the end of this step, you have two copies of the DNA strand of interest, and both return to their original helical form.

Polymerase

Polymerase attaches ....

... and extends

The Advanced Molecular Biology Lab at the University of British Columbia likes to play with and teach about DNA, RNA, and proteins. They publish the *Science Creative Quarterly* (scq.ubc.ca).

BACKYARD BIOLOGY

Hack Your Plants!

Play God in your garden — create custom fruits, flowers, veggies, and more. By Robert Luhn

Joe Real's backyard looks normal enough — a tidy rectangle of carefully trimmed grass with dozens of fruit trees lining the perimeter. But look closer: almost every tree looks like it fell down a flight of stairs, the limbs snapped off and replaced with new ones held in place with tape and rubber bands.

Welcome to the world of plant hacking. In Real's case, we're talking about grafting: the art of taking limbs (or even buds or chips) from one plant and suturing them onto another, to grow two or more varieties on the same plant.

One pear tree in Real's yard is growing 24 kinds of pears; another has two kinds of pears, two kinds of apples, and five varieties of quince hanging from its limbs. The centerpiece is a single citrus tree groaning under the weight of 48 different kinds of oranges, lemons, limes, and grapefruits. (The tree is currently being certified by Guinness World Records.)

METHODS AND MADNESS

Grafting's advantages are clear: it's a cheap, fast way to grow a cornucopia of plants in a small space. If you grow a tree from a seedling or sapling, you may have to wait four or five years for it to bear fruit; graft a stem from a mature tree, and you may get fruit next year. For gourmands like Real, it's a way to savor rare, mouth-watering fruit and vegetables year-round.

But grafting is a bit like collecting art. If you want to get more creative, the next step is crossbreeding — mating two different plants to create a unique offspring called a hybrid. With time, careful observation, and a bit of luck, you might produce a never-before-seen geranium; a newer, sweeter watermelon; or, alas, something god-awful. As one botanist put it, plant breeding is "a science that requires art."

HACKING THE HACKER

So why become a plant hacker?

"Everyone wants to put their hand on nature," says Todd Perkins, a flower breeder at Goldsmith Seeds. "It's primal. And there's the joy of taking infinite diversity and selecting just those traits you want."

Some of those traits might include creating a plant that blooms longer, can be harvested earlier, tastes better, is resistant to pests and disease, comes in groovy shapes and colors, or simply stands up straighter (a desirable trait in corn, for example).

But newbies and digerati, take note. This is no virtual experience — you're hacking life, sometimes literally, with a saw, and sometimes subtly, with a tiny brush coated with pollen. It's as tangible as life gets — we're talking dirt, bugs, plants, knives, saws, and, of course, sex.

P.S.: To get a handle on plant anatomy, check the handy diagram at makezine.com/go/anatomy.

IF YOU WANT TO GO DEEPER

We don't delve into techniques here that require fancy lab equipment, mutation-inducing chemicals, or a degree in molecular biology. For example, tissue culturing, mostly used for creating disease-free clones of a plant, can be used to merge individual cells from two distantly related species to create a new plant. If you want to take this plunge, check out the kits at kitchenculturekit.com.

We also don't wander into the realm of full-blown genetic engineering techniques. For that, you need more education and moola than called for here.

MAKING THE HACK

Grafting

With grafting, hack is the operative word. You basically splice a bud or a stem (called the *scion*) onto a plant with established roots (referred to as the *rootstock*). If you make a good fit between the active cells on the scion and rootstock (the *cambium*, the layer of cells right behind the bark) and you keep 'em in a moist, loving embrace, the two will eventually merge and form a solid connection in just a few weeks. The result: a new, strong branch that will grow a different fruit, flower, or whatever, on that plant.

Tools of the trade: **Grafting knife**, **grafting tape** such as Parafilm Grafting Tape, **anvil-style pruning shears**, a **hacksaw** (if you're grafting trees), **isopropyl alcohol** for sterilizing all blades before and after a graft, and a **wound sealant** such as Doc Farwell's Grafting Sealant.

Rules of the road: The plants you're melding should be closely related, or the graft won't take. Typically, you can graft plants from the same genus (such as members of *Prunus*: cherry, plum, apricot), and certainly from the same species. The rootstock should be about 2 years old — old enough to have an established root system, but young enough to still be growing like a teenager. The stem or bud you graft onto the rootstock should be dormant, which is why you typically graft in late winter or early spring.

Bud Graft

Hack the plant: There are dozens of grafting techniques, but we'll step through two of the classics: bud grafting and whip grafting.

A bud graft is the way to go if you don't have a lot of material to work with (like an entire stem) or you want to graft later in the year (midsummer and fall). Bud grafting works especially well with fruit trees and ornamental trees (such as magnolias).

First, select a stem that's about as thick as a pencil. Then, sterilize your knife with the isopropyl. About ¾ inch below a bud, cut ¼ inch into the stem at a 45-degree angle, and pull the knife back in a straight motion. Make a nice, flat cut, so the bud's cambium will cleanly make contact with the rootstock's cambium. When you cut the bud out of the stem, leave about ¾ inch of bark above the bud. You should end up with a bud sitting on a little rectangle of bark. Flip it over and remove any woody material.

Next, on the rootstock, take your knife and cut a "T" into the bark that's a bit bigger than the bud, and spread open the bark like little doors. Tuck the bud into place, as if you were slipping a foot into a shoe, and then let the flaps of bark cover it. Then, use grafting tape to secure the graft; wrap around and underneath the bud several times, and over the top once. (Don't put the tape on the bud.) Give the tape a good yank at the end, and it'll stretch and stick nicely to the rootstock. The graft is done! Finally, cut off anything above the bud graft on the rootstock; this will force the plant to devote more resources to the graft.

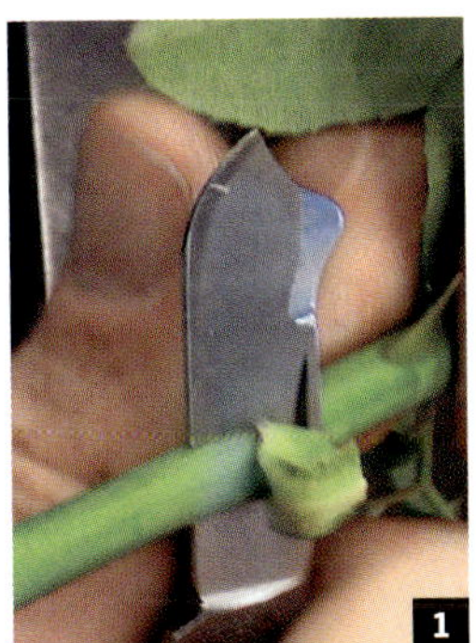

1. Remove a bud from a stem.
2. Flip over the bud and remove any woody tissue. The result should look like this — clean, green cambium.
3. Cut a "T" in the bark, pull back the bark (to reveal the cambium), and then slide the bud into the space (its cambium side down), like a foot into a slipper.
4. Wrap the graft: three times below the bud, once above.

Photography by Robert Luhn

Whip Graft

1

2

3

4

1. To make a whip graft, use a sloping cut at the bottom of the scion and a matching cut at the top of the rootstock.
2. Fit the two matching pieces together.
3. Wrap the graft with grafting tape.
4. Then apply grafting sealant to heal the wound and keep moisture in.

Hack the plant: A whip graft is a good choice if the rootstock (be it the trunk of a young tree or a major branch on an older tree) is about the same diameter as the scion you want to graft onto it. Whip grafts are strong and heal quickly. If you're new to grafting, start with apple or pear trees; they're simple to work with and forgiving of misaligned cambium.

First, cut off the top of the rootstock, leaving 6 to 12 inches. Remove any side shoots. For the scion, pick a stem about 9 inches long, cut just above a bud. The stem should be the same diameter as the rootstock, or slightly smaller. Make a 2-inch sloping cut at the end of the scion, and a matching cut at the top of the rootstock. On the cut surface of each, use your knife to create a little flap like a tongue. Then mate the scion and rootstock, fitting the flaps together. When the two cambium layers are kissing, wrap the graft tightly with tape, then apply a liberal amount of wound sealant all the way around the graft.

Attach a tag that notes the scion's species and the date of the graft. Keep a record in a ledger or on your computer, recording the rootstock, when the scion should bloom, the source of the scion, how well the graft took, how tasty the resulting fruit was, and so on.

TIPS, TRICKS, AND TRAPS

Grafting is pretty straightforward, even for someone who's all thumbs (like me). But plant pros (like the Plant Science professors at UC Davis) and avid amateurs (like Real) have learned some tough lessons and cool hacks:

Viruses be gone! Always buy scions that have been certified as free from disease.

Pick the right tools. Avoid scissor-style "bypass" shears, says UCD Professor Ali Almehdi. Buy sharp "anvil" shears that work by chopping. For the cleanest cut, make sure the blade's on top and the steel base is on the bottom. Get a true grafting knife made with carbon steel that's "flat ground," with the blade beveled on *one* side only, not both, for a cleaner cut.

If you can't match the cambium on both sides of the graft, just match one side and then wrap the graft. That should be good enough.

Don't leave the grafting tape on forever; it can impede growth. Remove it when a callus forms around the graft. Or open a wide rubber band so it's one long strip, and wrap it around the graft. The rubber will eventually deteriorate and fall off.

Keep your scions cool. "Dormant scion wood — stems from pear, apple, peach, cherry, and other trees — is best stored between 30 and 38 degrees [Celsius]," notes Real. The steps to success can be found online at makezine.com/07/graft.

Graft strategically. If you're smacking multiple grafts on a tree, think about placement. On his record-breaking citrus tree, Real grafted the most frost-resistant fruit at the top (mandarin oranges) and the least resistant at the bottom (limes and grapefruit). "The top level serves as a frost blanket for the citruses below," says Real.

By the same token, don't let a graft take over your tree. Graft vigorous scions on the northern side of the tree (less light means slower growth) and more mature scions on the southern side.

Graft the graft. Trying to graft distantly related plants (like apples and pears) and failing? Find an intermediary plant that's closely related to both, in this case quince.

MAKING THE HACK

Pollination

Tools of the trade: Tweezers, small clean paintbrush, magnifying glass, glassine bag

Rules of the road: Being the first on your block with a DaffoMelon would be cool, but don't hold your breath. Typically, you can cross plants of the same species to create a new variety. You can also make successful crosses between different species in the same genus, such as the raspberry (*Rubus idaeus*) and blackberry (*Rubus rosaceae*), which results in a loganberry. Rare, however, is a cross between plants from two different genera. Adjust your expectations accordingly.

As you might expect, crossbreeding is easiest with plants that boast big sex organs: clearly defined and separated stamens and pistils. Two obvious candidates would be daffodils and tulips, but get the MO on a plant's life cycle before you cross it with anything. You might have to wait several years before a flower appears. (Our reference of choice: Charles Welch's *Breeding New Plants and Flowers*.)

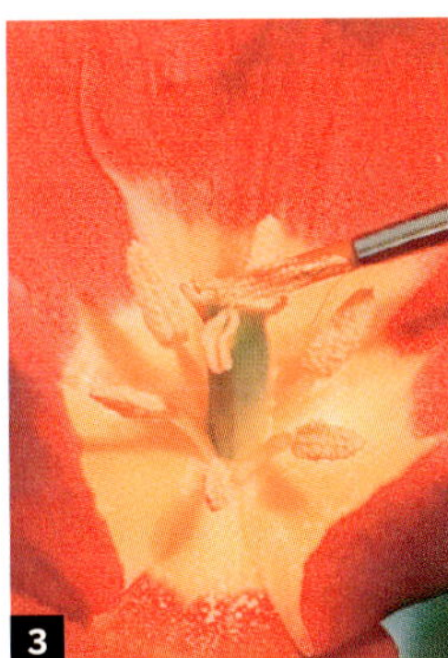

1. Remove the stamens to prevent the flower from self-pollinating.
2. Gather pollen from the donating flower with a small paintbrush and bring it to the target flower.
3. Brush pollen against the flower's stigma. The pollen will migrate down the tube, fertilize the flower's eggs, and start making seeds.

Hack the plant: To get the 411 on the mating dance, we spent the day with Todd Perkins, a flower breeder at Goldsmith Seeds. His recommendation? First, work with a plant that normally has prominent organs (such as the petunia), so you can easily control the pollination. (If you don't, you'll definitely need that magnifying glass.) If you have the time, grow several varieties of the plant and let them cross naturally. Select the offspring you like, and grow their seeds. Repeat this process until you have two plants with the traits you want to meld.

Using tweezers, emasculate (ouch) the target flower by removing its stamen. (The stamen often look like tiny stalks that ring the center of the flower; the pollen-bearing tip is called the anther.) This will prevent the plant from pollinating itself.

From the source flower, collect pollen from the tip of the stamen with a small paintbrush. Trot over to the target flower, locate its stigma (it's often a tall tube in the center of the flower, which leads to the plant's ovaries) and brush the pollen on it. Tag the flower that's been pollinated, noting date and which plant/flower was the pollen source. Finally, slap a chastity belt — a glassine bag — over the flower to prevent pollination from other sources.

In less than 24 hours, fertilization will occur and the flower will wilt. (Why? First, to prevent further pollination, and second, to force the plant to marshal its resources for reproduction.) In as little as four weeks, you'll get seeds. Plant the seeds and you'll get your hybrid plant, right?

Not exactly. Thanks to the nature of sexual reproduction and recombination, you'll get plants exhibiting a huge mix of characteristics from their grandparents.

One solution? From this first batch of plants, pick the ones that are closest to your ideal, and mate

Photography by Mark Yates

them with the best parent. (A practice known as *backcrossing*.) This will produce seedlings that contain 75% of the parent's genes. Pick the plants from this next batch and pollinate each plant with its own pollen. This should effectively lock in the characteristics you want. The seeds from these plants should be nearly identical and generate the hybrid you want.

The big exception? This only works if you're selecting for *simply inherited* traits — those tied to a single gene. With traits controlled by multiple genes, the plants you grow from seed will still be all over the map.

The second solution? Once you've got a plant that's close to your ideal, clone it. Take cuttings from the plant, place them in the appropriate medium (such as peat and perlite), add water and perhaps a rooting hormone, and soon shoots and roots will develop. Plant 'em and voilà — up comes your hybrid. "This is the lockbox," says Goldsmith's Perkins. "It's the one sure way to preserve the hybrid that you've created." (For more on cuttings, see makezine.com/go/cuttings.)

THEM THAT PLANTS ...

As you've probably guessed, this article just scratches the soil on the topic of plant hacking. There are lots of different techniques and lots of variation among plants. (You'll discover, for example, that some plants stubbornly resist crossbreeding and that anatomy varies wildly from species to species.) And, of course, there are many related issues to explore at length, from plant care and feeding to propagation.

The biggest challenge, however, is mental.

"There's a fine line between gardening and madness," says Andy Mariani of Andy's Orchard in Morgan Hill, Calif. But, as obsessions go, plant hacking is one of the better ones.

Special thanks to Dr. Tom Gradziel and Dr. Ali Almehdi of UC Davis; Todd Perkins of Goldsmith Seeds; Andy Mariani of Andy's Orchard; Dr. Carol M. Stiff of Kitchen Culture Kits; and Lisa Stapleton of the California Rare Fruit Growers association.

Robert Luhn is a technology and science writer based in El Cerrito, Calif., who spends his free time raising unattractive plants.

TIPS, TRICKS, AND TRAPS

Just starting out? Limit yourself to one or two projects. Good starter veggies and fruit include tomatoes, squash, and melons. Good starter flowers include geraniums, fuchsias, and petunias.

Pollen has a limited shelf life. You've got 24 to 48 hours, tops. So check the donating flower as soon as it opens. With many plants, fresh pollen is soft, loose, and fluffy, and comes off the anther easily.

Get Mendelian. "This isn't for the casual gardener," says Andy Mariani. "To juggle the recessive and dominant traits, you must have an elementary understanding of genetics." Start with the books listed in the What to Read section below.

Pick the right parents. Closely examine the two plants you want to cross. Are they robust, superior specimens, from buds to color to flowering? Only pick the best of the best.

Nip it in the bud: Stop self-pollination at all costs. Gently open the flower while it's still a bud and remove the stamens. Then cover it for a day or two, until the stigma is ready to take the pollen.

When it comes to sex, timing is everything. The pollen from the source flower must be fresh and the stigma in the receiving flower must be ready. How can you tell? This varies by species, but in some flowers, the stigma grows tall and splits — a sure sign it's ready for pollen. In others, the stigma becomes sticky and glistens (the mucus captures any pollen blowin' in the wind or dropped by a passing bee). If the stigma is dry, the moment is gone. NOTE: Some pollen is designed to be sticky, so bees will carry it off.

WHAT TO READ

***Breeding New Plants and Flowers*, by Charles W. Welch (Crowood, 2002)**

***Breed Your Own Vegetable Varieties: The Gardener's and Farmer's Guide to Plant Breeding and Seed Saving*, by Carol Deppe (Chelsea Green, 2000)**

***Plant Propagation, The American Horticultural Society*, Alan Toogood, editor in chief (DK, 1999)**

Make: Projects

Prepare for launch with a reusable rocket cam that records the ups and downs of rocket life. Then crank up your crankshaft and feel the heat of a two-can Stirling engine. Or watch your new hobby mushroom into an obsession by growing fungi in your own mini clean room.

ROCKET-LAUNCHED CAMCORDER

By John Maushammer

Photography by John Maushammer

EYES ON THE RISE

Hack a $30, single-use camcorder to make it reusable, then launch it up in a model rocket and capture thrilling astronaut's-view footage of high-speed neighborhood escape and re-entry.

You can build this project over a weekend and the results are fantastic. The idea goes back to 1929, when Robert Goddard launched the first scientific payload on a rocket: a still camera and a barometer. During the height of the space race, model rocketry supplier Estes offered a tiny Super 8 film kit that recorded about 10 seconds of rocket POV action. Today, Estes sells a launchable DV recorder called The Oracle, but it costs $120 and its image quality is lacking.

We'll do much better with a new camcorder that costs only $30, is very light, and has enough memory for several flights. It was designed for single-use only, but we'll make it reusable.

NOTE: Pure Digital has changed the firmware in later versions of the camera, which may make your images impossible to download. Check the current states of hackability in the forums at camerahacking.com, or check other options at makezine.com/07/camerarocket.

Set up: p.81 **Make it:** p.82 **Use it:** p.89

John Maushammer reverse-engineered the firmware in all three Pure Digital disposable cameras and figured out how to connect them to home computers. While not technically a rocket scientist, he has designed hardware and software for satellites.

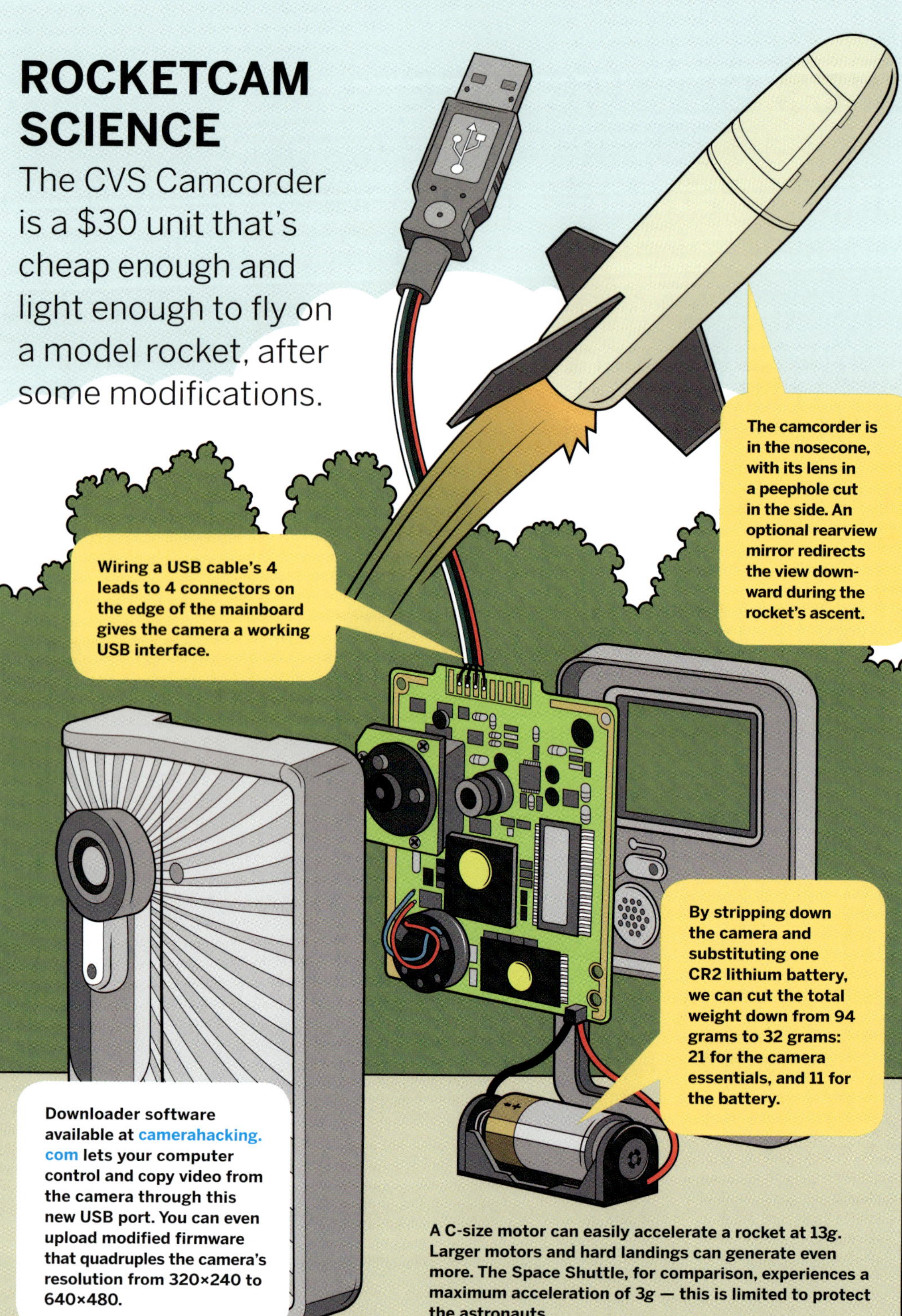

Illustration by Timmy Kucynda

SET UP.

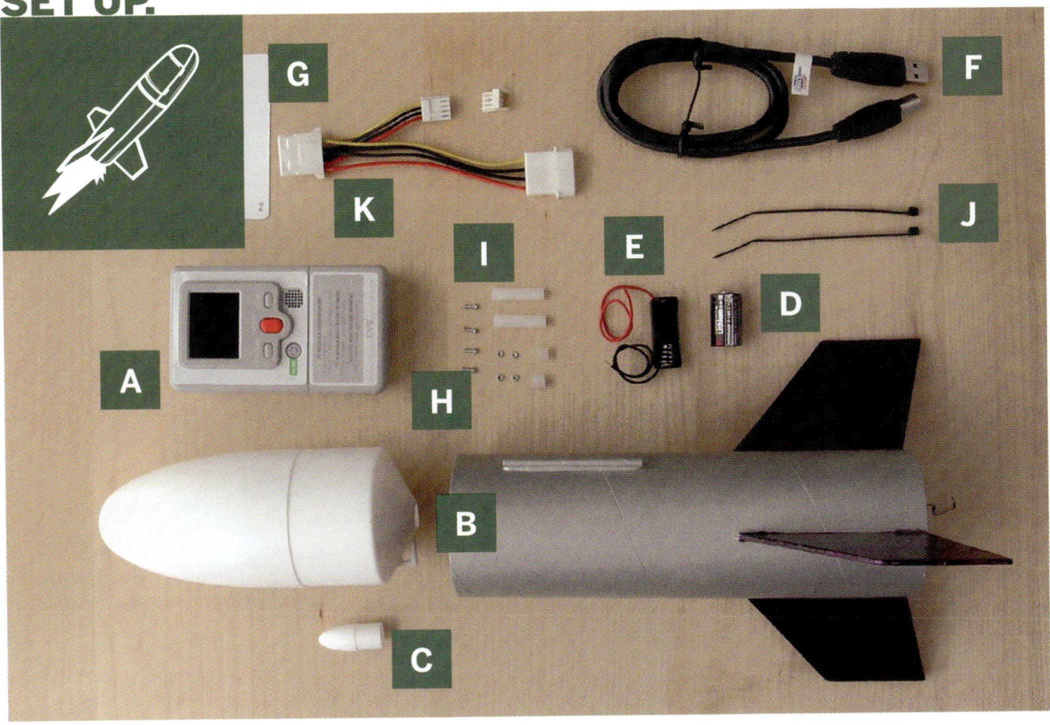

MATERIALS

[A] CVS One-Time-Use Video Camcorder $30 at CVS drugstores. Rite Aid sells a similar camera; check camerahacking.com for compatible drivers. Target carries a reusable Point and Shoot Video Camcorder (no hacking required), but it costs $130.

[B] Model rocket kit The body tube must be at least 2½" in diameter. Also, check the Estes Engine Chart (*see Resources*) to make sure the engine can lift 41 grams of extra weight. Two recommended rockets are the Fat Boy and the Canadian Arrow, both from Estes.

[C] Small rocket nose cone (optional) Estes sells spare nosecones in packs.

[D] Small 3-volt lithium battery Such as a CR2

[E] Battery holder You can adapt an N-sized holder to fit a CR2.

[F] USB cable

[G] Credit card plastic An old credit or gift card, or fake card from junk mail

[H] Small machine screws and matching nuts (4) No. 1 or M1.8 metric

[I] Nylon standoffs big enough to glue the nuts into (4) Or threaded standoffs that match the screws, if available

[J] Small plastic cable ties (2)

[K] A mating pair of lightweight, 4-pin free-hanging power connectors I used connectors from inside an old floppy disk.

Thin wire

Small front-surface mirror (optional) You can find a good one inside a View-Master toy ($5), but be sure to get the kind that looks like binoculars.

Masking tape

Engines and igniters With the smaller Fat Boy rocket, you'll need to upgrade to a C11 or D-sized engine and use a larger motor mount. The Canadian Arrow's standard D or E engine has enough power in stock configuration.

Launch controller Such as Estes' E controller (30' cable) or Electron Beam controller (17' cable)

Parachute recovery wadding Heat-resistant paper that prevents the parachute from melting.

Launch pad (stand, blast shield, and guide rod) Also from Estes (*or see MAKE, Volume 05, page 141, for a DIY version*)

TOOLS

Windows XP computer

Small Phillips screwdriver

Keyhole saw or hacksaw

Hobby knife, scissors

Soldering equipment

Wire cutters and stripper

Polyurethane glue, such as Elmer's ProBond or Gorilla Glue This glue is tough and foams up to fill gaps.

C-clamps (2)

Vice Helpful to hold parts while cutting or sawing.

MAKE IT.

BUILD YOUR CAMCORDER ROCKET

START **Time: A Weekend** **Complexity: Medium to Easy**

1. BUILD THE ROCKET BODY

1a. Follow the instructions to assemble your model rocket kit.

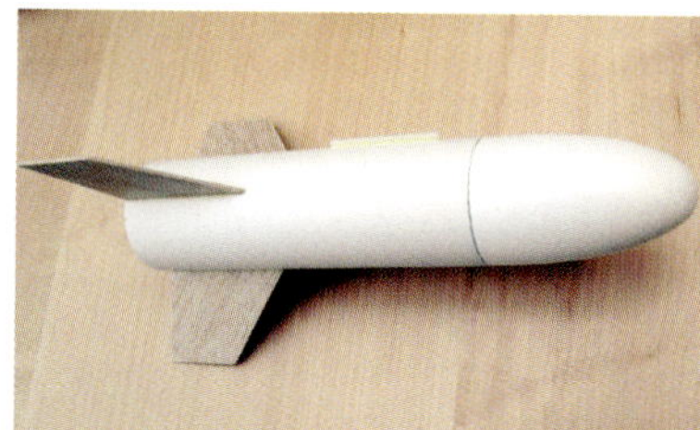

After any gluing and painting steps, you can skip ahead and work on installing the camcorder in the nosecone.

2. DISASSEMBLE AND STRIP DOWN THE CAMCORDER

The CVS camcorder is already small, but it's still too heavy to fly in most rockets. By removing everything that isn't essential, we can cut its weight down from 140 grams to 21 grams (without batteries).

2a. Remove the camcorder's battery cover by inserting something pointy into the opening on the bottom while sliding the cover off.

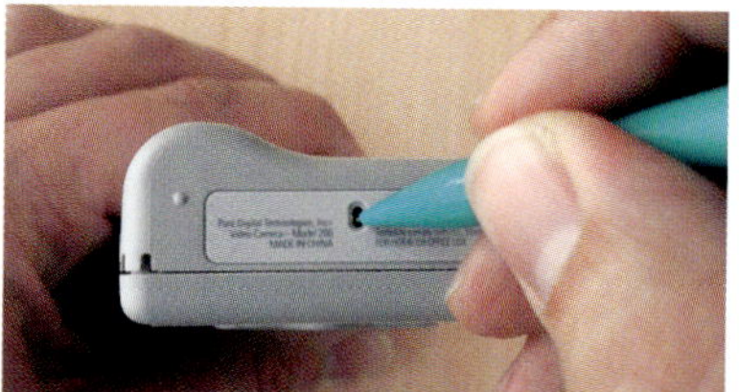

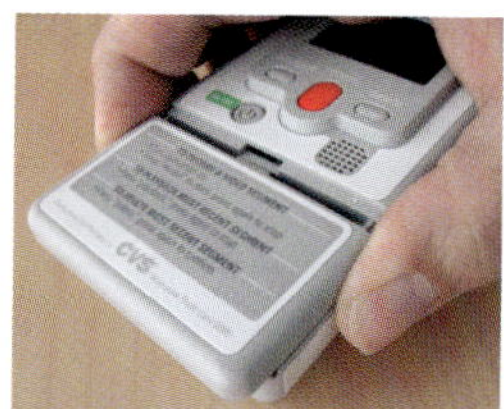

2b. Unlock the grey battery holder and remove it.

2c. Unpeel the sticker on the back and use a small Phillips screwdriver to remove 4 screws, one at each corner.

2d. Snap apart the case. Inside is the main circuit board.

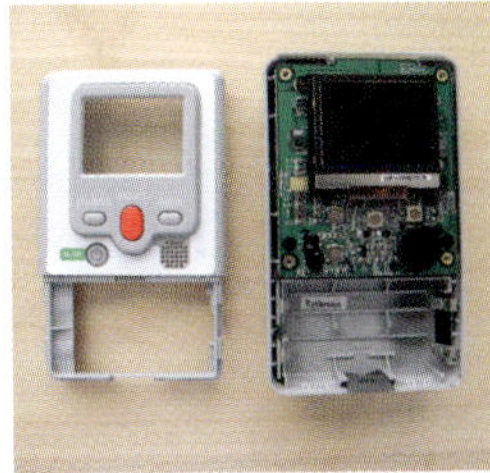

2e. Remove the 2 screws at the 2 points shown here to release the circuit board.

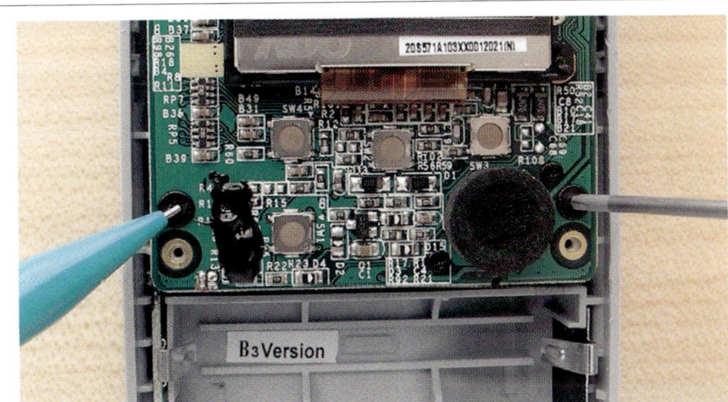

2f. Find the small 4-pin connector that connects the circuit board to the batteries. Pull straight up to remove the board.

2g. Optional: Remove the speaker, which is encased in a vibration-resistant rubber housing glued to the circuit board. Just pull it to remove, and cut or unsolder the speaker wires.

Removing the speaker saves 2 grams, which doesn't seem like much, but every bit counts on smaller rockets. Leave the microphone in place, though, to record the roar of takeoff.

3. RIG THE CAMERA INTERFACE

The camera connects to a computer using USB protocol, but instead of a standard USB port, it has its own card-edge connector that we'll need to wire into. We'll save flying weight by using 2 USB cables, a short one that's attached to the camera and ends in a lightweight connector, and a longer, second cable that connects the lightweight connector to your computer's USB port.

3a. Cut and strip a short section of the USB cable. Solder the red, black, green, and white leads to the mainboard's edge connector, contacts 6-9, respectively. (If the computer indicates a problem later, your cable may not have the standard color-coding, and you should try swapping the green and white wires.)

This cable makes the same connections as the cable for the Dakota Digital camera in MAKE, Volume 03 (*"Cheap Shot," by Charles Hoffmeyer, page 130*).

3b. Solder or crimp the 4 leads to one of your power connector pairs, and solder or crimp the other connector to the computer end of the USB cable, preserving the wire ordering.

My connectors weren't designed to be taken apart that often, but shaving down the locking tabs made them disconnect more easily.

3c. Plug the cable into the computer. Windows XP should identify it as a "Saturn" and prompt you to install drivers. These drivers don't exist yet, so hit "Cancel."

If the computer gives an error message and cannot identify the camera as a "Saturn," swap the green and white USB connections.

3d. Browse to camerahacking.com. Under "CVS One-Time-Use Camcorder," click "FAQs & Links," and choose some up-to-date driver software. Carpespasm, BillW, and Corscaria have good wares, among others (and I wrote a Mac downloader that works with the oldest version of the camera, 3.40). Install per instructions.

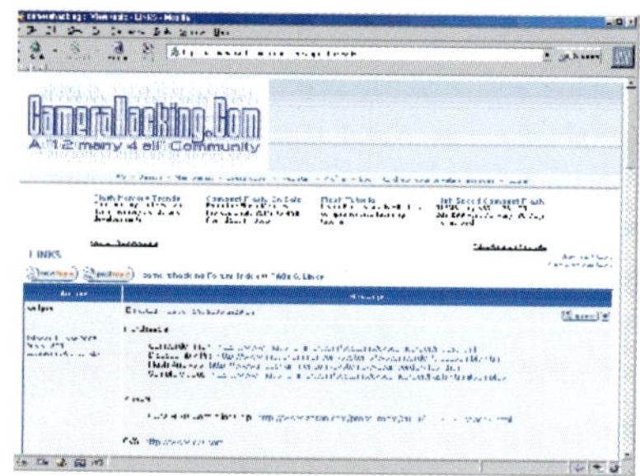

3e. Record a video and test it. It will be in XVID 1.0 format, 320x240 resolution, 30 frames/sec. After you're able to download videos from the camera, you may need to install a video codec to play them back; I recommend MPlayer from mplayerhq.hu.

4. MOUNT THE CAMERA IN THE NOSECONE

Now we'll put the camcorder in the nosecone. This is the most protective part of the rocket, and adding the weight in the front will help stability during flight. There are many ways to secure a camera so it won't come off during liftoff, including styrofoam and glue, but this method is reversible, which I prefer.

4a. Cut a hatch in the side of the nosecone, for inserting the camera. I taped a 3"×2¾" piece of paper to the cone to use as a template. We will be replacing this hatch, so don't cut any more than necessary.

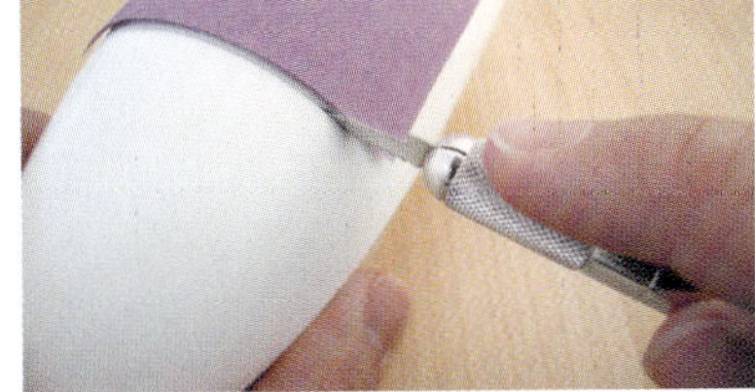

The hatch should be big enough to let you operate the camera. Don't cut all the way to the bottom of the cone; leave a ring above the base to keep it strong. If you mess up, you can buy a replacement cone.

4b. Make 4 threaded standoffs by gluing M1.8 nuts into the unthreaded standoffs. (The circuit board's mounting holes are so small that I couldn't find threaded nylon standoffs that would fit.) Screw the standoffs onto the board.

Score the inside of each standoff to give the glue something to attach to. Make sure the glue is fully cured before screwing the standoffs on.

4c. Cut the standoffs so that they match the curve of the nosecone. Start with 2 short and 2 long. Use the rocket body to mark them, and cut them more precisely, angled to fit.

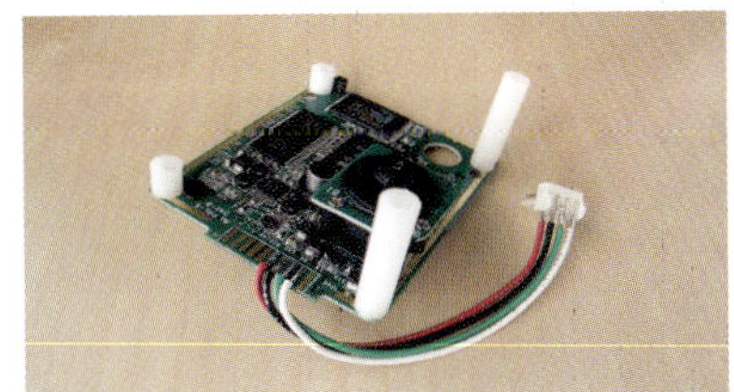

4d. Solder the battery holder's negative lead (black) across the 2 pins closest to the lower-left corner, near the on/off button. Solder the positive (red) lead to bridge the adjacent 2 pins, near the mystery-chip blob. If you're using an N battery holder with a CR2 battery, cut the sides off of the holder to let the battery seat. CR2s are the same length as Ns, but wider.

4e. Cut a peephole in the side of the nosecone and verify it with the camera's viewfinder. The closer you can mount the lens to this hole, the smaller it can be. But don't worry about the size yet — you can enlarge it after the camera is glued into place.

4f. Verify that everything fits, the camera lens aligns with the peephole, and there's enough wire to reach the battery holder, which you'll mount below. Mark where the standoffs will glue to inside the nosecone, and rough these spots up. I used a knife to cut a crosshatched pattern.

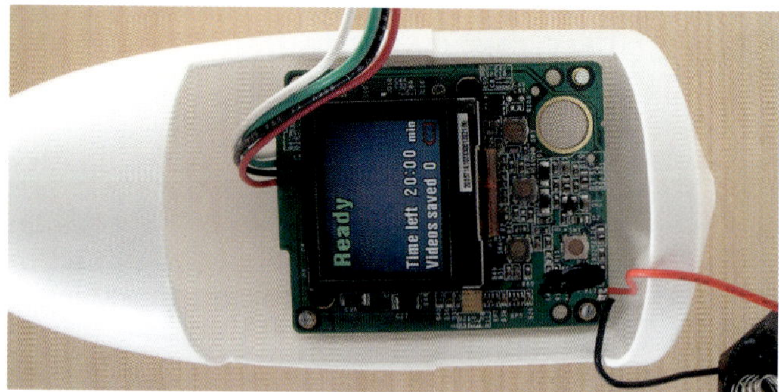

4g. Glue in the standoffs and double-check that the camera can still see out the hole. Use C-clamps to hold the board in place while the glue foams up. Allow to cure completely. Polyurethane glue requires moisture to cure, so I put a small wet paper towel in the cone.

4h. After the initial application of glue cures, add some more to reinforce the standoffs.

4i. Optional: If you want a view of the ground as the rocket lifts off, install a rearview mirror on a small nosecone, attached in front of the peephole. The reflection will make the body of the rocket appear at the top of the picture. If you want the body at the bottom of the picture, you can mount the camera upside-down, or correct the videos digitally later. Cut a plastic View-Master mirror down to size with wire cutters. After the glue has set, trim around the mirror with a hobby knife and file.

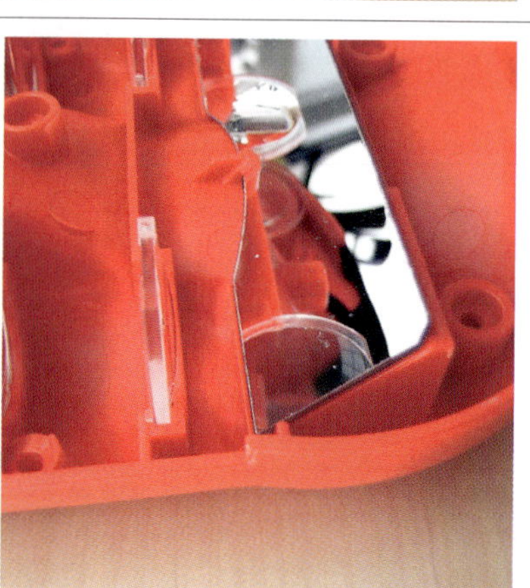

View-Masters (and SLR cameras) use front surface mirrors, where the coating is on the front of the plastic (or glass). This prevents image ghosting, but the shiny coating is easier to scratch.

4j. Optional: To save even more weight, remove the LCD viewfinder. Remove the camera circuit board from the nosecone and unscrew the 2 screws that hold the screen on the back. Slide the gray plastic catch on the screen's ribbon cable (I used 2 small screwdrivers to help), and pull to completely disconnect the screen and cable.

5. INSTALL THE BATTERY

Battery holders have no springs on the positive terminal, so a good whack upon landing can pop the battery out, interrupting the current. If this happens before you press the Stop Record button, you may write garbage to the camera's memory chip and lose your entire recording, or even render the camera inoperable. To prevent this, I strung a small wire through the positive terminal of the battery and connected it to the positive terminal of the battery holder. That way, if the battery is knocked loose, the wire still keeps it connected electrically.

5a. Thread some thin wire through the vent holes on the positive terminal of the battery. If you don't have wire small enough, use a few strands from a larger, stranded wire.

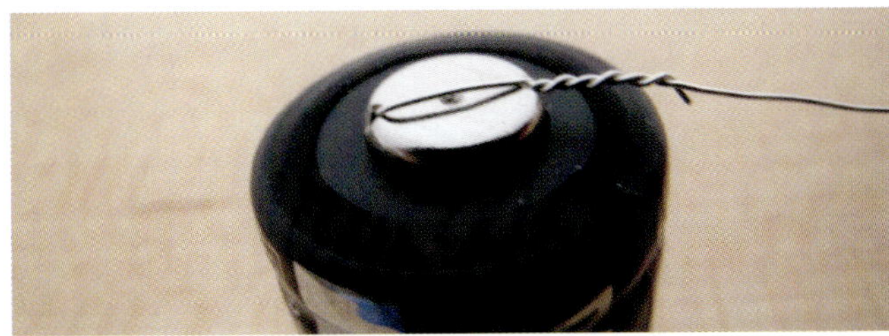

5b. Load the battery into the holder and connect the wire to its positive terminal. Glue the holder into the bottom of the nosecone, below the hatch opening.

5c. Cut two small holes through the bottom of the cone and thread a cable tie through the holes and over the battery, to hold the battery in place. Because the battery is mounted perpendicular to the direction of flight, the cable tie will absorb most of the shocks.

6. FINISH THE NOSECONE

6a. Referring to the camera's original housing, label the Power and Record buttons on the bare circuit board so they are easy to use.

6b. Add some tabs to be able to close the hatch securely. Glue 2 thin strips of credit card plastic inside the cone along the sides of the hatch. Glue 2 smaller pieces to the front of the hatch itself, and a third small piece inside the nosecone, to fit in between. When done, the hatch should sit flush with the nosecone.

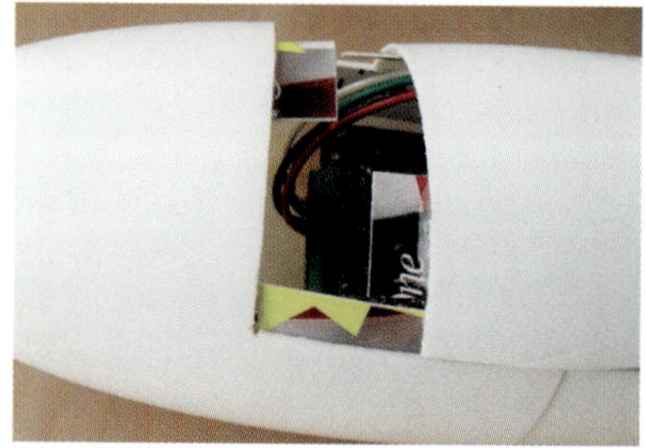

Make sure you reinforce the hatch at the bottom, to keep it securely in the rocket, and at the leading edge, where it will bear the most pressure.

6c. Plug any holes in the bottom of the nosecone by gluing more pieces of credit card plastic.

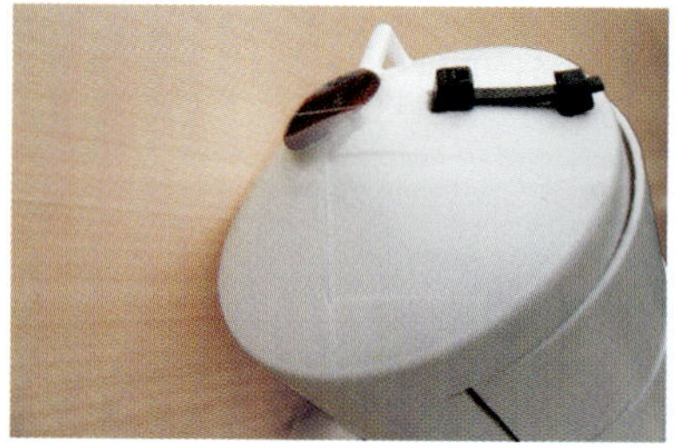

At the top of the rocket's flight, a small explosive charge fires in the rocket motor to pop the nose off and deploy the parachute. Holes in the bottom of the nosecone can vent pressure without popping the cone, which will send your rocket crashing into the ground!

FINISH X

NOW GO USE IT »

Rocket Stability

Did you ever wonder why rocket fins stick down past the tail? The reason has to do with stability, which you need to understand if you are modifying a rocket kit.

Two imaginary points determine whether a rocket will fly straight or corkscrew hopelessly out of control. The first point is the *center of gravity* (COG). If you were to place your rocket on a razor's edge, this is the point where it would balance. Add weight to the nosecone and the center of gravity moves toward the nose; add bigger engines and it moves rearward.

The second point is called the *center of pressure* (COP), and it is a little harder to explain. This is similar to center of gravity, except it involves the aerodynamic forces balancing out. If you were to trace the 2-dimensional side-view of the rocket onto cardboard and then cut it out, it would balance near the center of pressure. Enlarging the fins and extending them down moves this point rearward.

Once you know these two points, you can estimate the rocket's stability. The general rule for stable flight is that the COG must be at least one body diameter in front of the COP. During flight, aerodynamic forces will push rearward on the COP. If it isn't already behind the COG, then the rocket will attempt to turn around, which is bad news. If the COP is behind the COG by just a little, the flight will be marginally stable and probably corkscrew.

For this project, we won't be modifying the outside too much, so the COP will remain the same. We will be adding the weight of the camcorder to the nosecone, moving the COG forward. Aerodynamically, this will only make the flight more stable.

USE IT.

NOW GIVE IT A SHOT

LAUNCH SEQUENCE

If you are using a smaller engine, you may want to do a final weigh-in before going ahead with a launch. With everything installed, verify that the total weight is within the capability of the rocket motor. If so, follow the rocket kit's launch instructions. Here's the basic sequence:

1. Check to make sure that the nose fits securely, but not too tightly. Then tie up the parachute, engine, and wadding.
2. Set up the launch pad per instructions. Make sure the field you are launching in is large enough; otherwise you'll lose your precious payload. For a rocket powered by a D-sized engine, the field should be at least 500 feet in diameter.
3. Ensure that the rocket launcher is not armed (usually this means the key is removed) and then set up the rocket on the pad.
4. Turn on the camera and press the record button. The red record light, just under the lens, should come on.
5. Close the hatch and tape it closed. Although not typically rated for space-faring use, ordinary masking tape works fine.
6. Start the countdown and launch.
7. Recover the rocket. Open the hatch and press the Record button to stop recording. The recording light should turn off. Then turn the camera off by pressing the On/Off switch. If you forget to turn the camera off, it will switch off automatically after a few minutes of non-use. Do not turn the camera off by removing the battery.
8. Back at your computer, hook up the USB cable, download the video, and enjoy.

WATCH IT

Watch a high-flying video captured by John Maushammer's Rocket-Launched Camcorder at makezine.com/07/camerarocket.

UPGRADES

Downward View During Descent

If you installed the downward-facing mirror, you'll get a whole new view of the launch. One drawback, though, is that the descent will typically have views of the parachute and sky. You can change this by attaching the parachute to the tip of the nosecone instead of the base. Add a small eyehook to the nosecone's tip, and run the parachute cord along the side of the nose and into the body.

Improved Resolution

The CVS camera (which is manufactured by Pure Digital, along with the Rite Aid and Target cameras) actually has a 640x480 sensor. But in order to extend recording time, it is configured to record at only one quarter of this resolution. Recording time isn't a problem for rocket flights that last only a few seconds, so you can set the camera to record at the full resolution. You can do this by uploading a modified version of the binary file *USP.BIN* into the *P3* directory of the camera. See Resources, below.

RESOURCES

Estes Engine Chart
makezine.com/go/estes
Determining Center of Pressure
makezine.com/go/cop
Downloader software
makezine.com/go/software
Improving image resolution
makezine.com/go/resolution

TWO-CAN STIRLING ENGINE

By William Gurstelle

Photograph by Kirk von Rohr

REDLINING AT 20 RPM

The Stirling engine has long captivated inventors and dreamers. Here are complete plans for building and operating a two-cylinder model that runs on almost any high-temperature heat source.

Stirling engines are *external combustion* engines, which means no combustion takes place inside the engine and there's no need for intake or exhaust valves. As a result, Stirling engines are smooth-running and exceptionally quiet.

Because the Stirling cycle uses an external heat source, it can be run on whatever is available that makes heat — anything from hydrogen to solar energy to gasoline.

Our Stirling engine consists of two pistons immersed in two cans of water. One can contains hot water and the other cold. The temperature difference between the two sides causes the engine to run. The difference in the hot and cold side temperatures creates variations in air pressure and volume inside the engine. These pressure differences rotate a system of inertial weights and mechanical linkages, which in turn control the pressure and volume of the air cylinder.

Set up: p.94 **Make it:** p.96 **Use it:** p.101

William Gurstelle serves on MAKE's Technical Advisory Board and is the author of *Backyard Ballistics* and *Adventures from the Technology Underground*.

THE STIRLING CYCLE

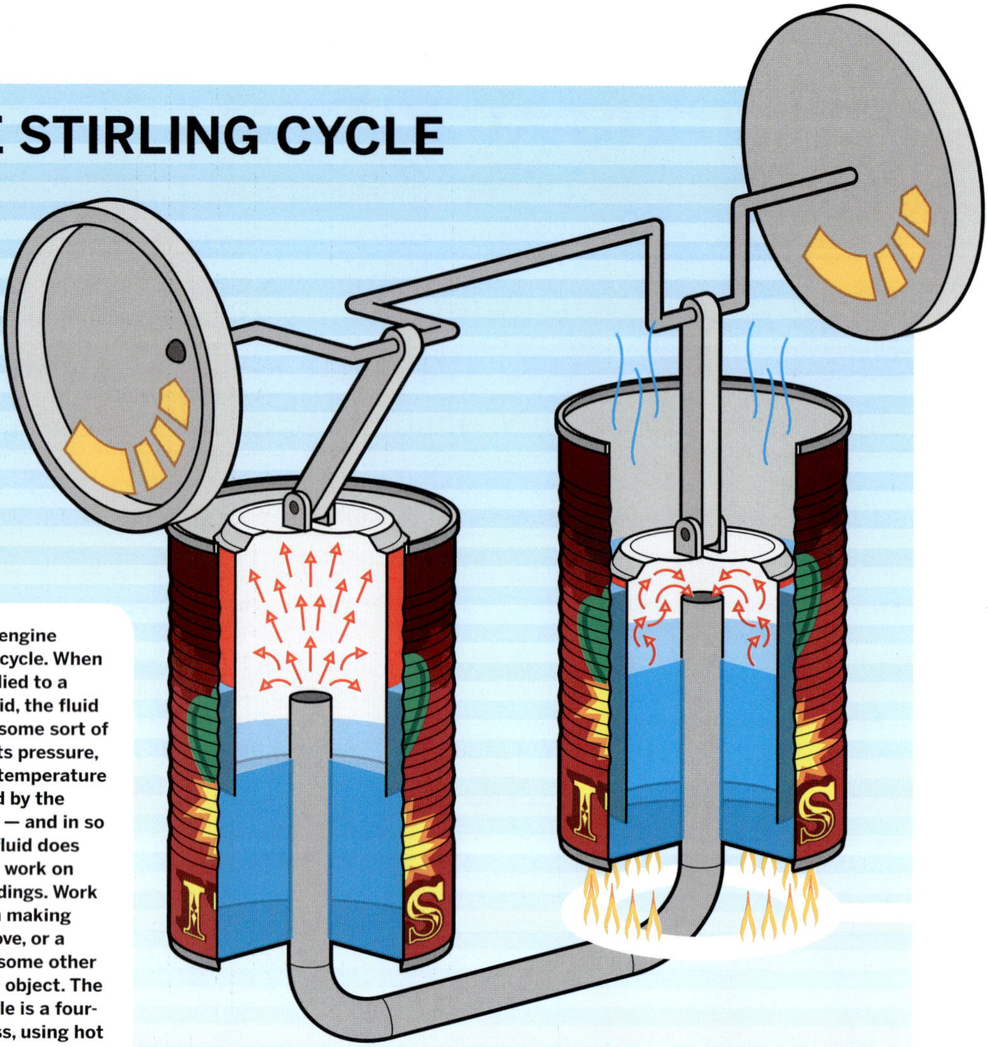

Every heat engine works on a cycle. When heat is applied to a working fluid, the fluid undergoes some sort of change — its pressure, volume, or temperature is increased by the added heat — and in so doing, the fluid does meaningful work on its surroundings. Work could mean making a piston move, or a turbine, or some other mechanical object. The Stirling cycle is a four-step process, using hot air as its working fluid.

Four Steps of the Stirling Cycle

1

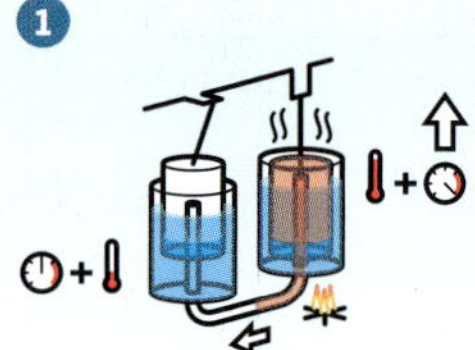

COOLING

Cold piston (left) moves upward by flywheel inertia, drawing hot air over to cold side.

2

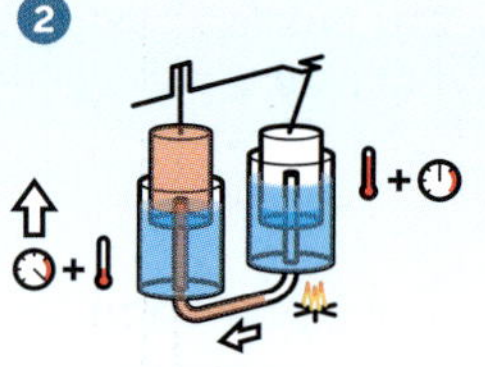

EXPANSION

Hot air is forced to the left cylinder, forcing the cold piston up. This is the power cycle.

3

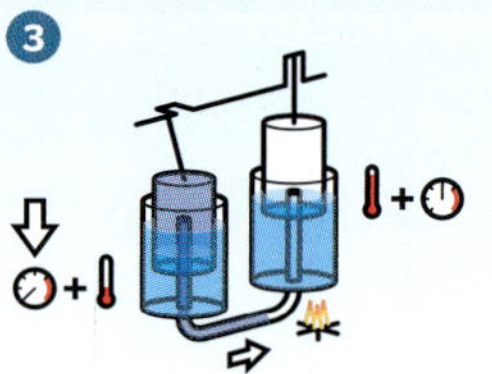

COMPRESSION

As air in the cold water contracts, the cold piston moves down.

4

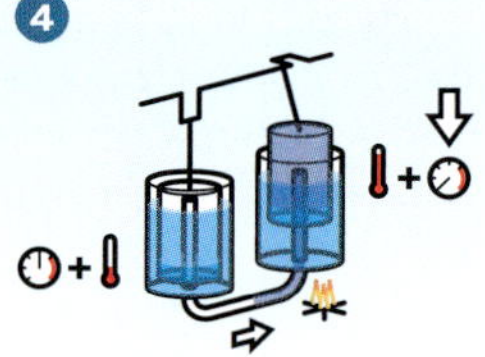

HEATING

With the cold piston fully down, most air is on hot side and getting reheated.

Illustrations by Nik Schulz/L-Dopa.com

THE STORY OF THE STIRLING ENGINE

All engines run on heat cycles. More properly called thermodynamic cycles, each of these cycles has a name. Cars run on the Otto cycle, trucks on the Diesel cycle. Power plants often run the Rankine, while gas turbines run the Brayton cycle.

One cycle in particular has long captivated inventors and dreamers — the Stirling cycle. The Stirling cycle was among the first of the thermodynamic cycles to be exploited by engineers. Compared to other engine types, it is ancient. When it was patented as a new type of engine by a Scottish cleric in 1816, scientists hadn't even come up with the idea of thermodynamic cycles.

Robert Stirling, a young Scottish Presbyterian assistant minister, had the idea for a new type of heat engine that used hot air for its working fluid. Until then, the steam engines of Watt and Newcomen were the only heat engines in use.

Stirling Engines Go to Work ... and Are Laid to Rest

Stirling's idea was to alternately heat and cool air in a cylinder using articulated mechanical arms and a flywheel to coax the machine to run in a smooth, endless cycle.

Although complex and expensive for its time, Reverend Stirling made it work. As early as 1818, his engine was in use pumping water from a stone quarry. By 1820, a 45-horsepower Stirling engine was driving equipment in the Scottish foundry where his brother worked.

Auto manufacturers have experimented with the Stirling for years. Its numerous good qualities make the Stirling an attractive candidate to replace or augment internal combustion engines.

Automakers worked closely with the federal government from 1978 to 1987 on Stirling engine programs. The goals were ambitious: low emission levels, smooth operation, a 30% improvement in fuel economy, and successful integration and operation in a representative U.S. automobile.

General Motors placed one in a 1985 Chevrolet Celebrity, and met all of the program's technical goals. But improvements in the efficiency of existing engine types, coupled with the status quo's far less expensive cost structure, doomed the Stirling to automotive irrelevance.

The External Combustion Revival

The Stirling idea was dusted off in the mid-1990s. A prototype Stirling hybrid propulsion system was integrated into a 1995 Chevrolet Lumina. But that test was not particularly successful, as the hybrid vehicle failed to meet several key goals for fuel efficiency and reliability. The program was abandoned. Still, Stirling engine advocates continue to research and apply the technology. The big breakthrough may yet arrive, possibly in a hybrid electric-Stirling engine.

While not terribly complex, the engineering analysis of the engine's thermodynamic cycle goes beyond the scope of this article. Suffice it to say that Stirling engines operate on a four-part cycle in which the air inside the engine is cyclically compressed, heated, expanded, and cooled, and as this occurs, the engine produces useful work.

While most heat engines are fairly understandable to interested amateurs, building one yourself is an altogether different prospect. Most engines require carefully machined metal parts, with close tolerances and tightly sealing clearances for pistons and/or rotating parts. Robert Stirling's heat engine is an exception. Or at the very least, making a working model can be done without any difficult machining.

About MAKE's Stirling Engine

This article provides step-by-step instructions for building a straightforward Stirling external combustion engine.

This engine is simple and cheap, and once you get it going, you really get a feel for how this sort of engine works. It chugs along at a leisurely 20 to 30 rpm, its power output is minuscule, and it makes a delightful squishing/chuffing noise as it operates.

But be forewarned: All engines, even the metal-can Stirling described here, are complex mechanical devices in which myriad mechanical movements must come together in precise fashion in order to attain cyclical operation.

SET UP.

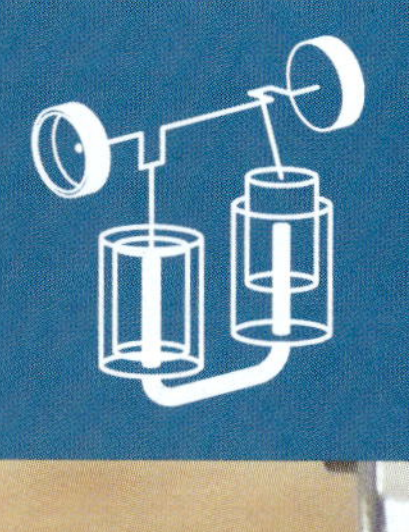

MATERIALS

Large steel cans (2) At least 4" in diameter. Large juice cans or 1lb. coffee cans work; 13 oz. coffee cans are too small.

Copper gauze Such as "Chore Boy" pot scrubber

Aluminum soda cans (2)

#3 size rubber stopper To fit middle opening of the copper tee

Plastic spacers, 1" long (2) The spacer's outside diameter must match the inside diameter of the sheave, while its inside diameter must just fit the rod used for the crank. Look in hardware stores, in the small parts bins that contain specialized fasteners.

¾" copper tee

¾" copper pipe, about 18" long Cut as follows: 2¾" (2), 5" (2)

5"-diameter metal die-cast sheaves or pulleys (2) Such as McMaster-Carr #6245K45

Wood 1"×2", 9" long (2) Pieces A

Wood 1"×10", 10" long Piece B

Photograph by Kirk von Rohr

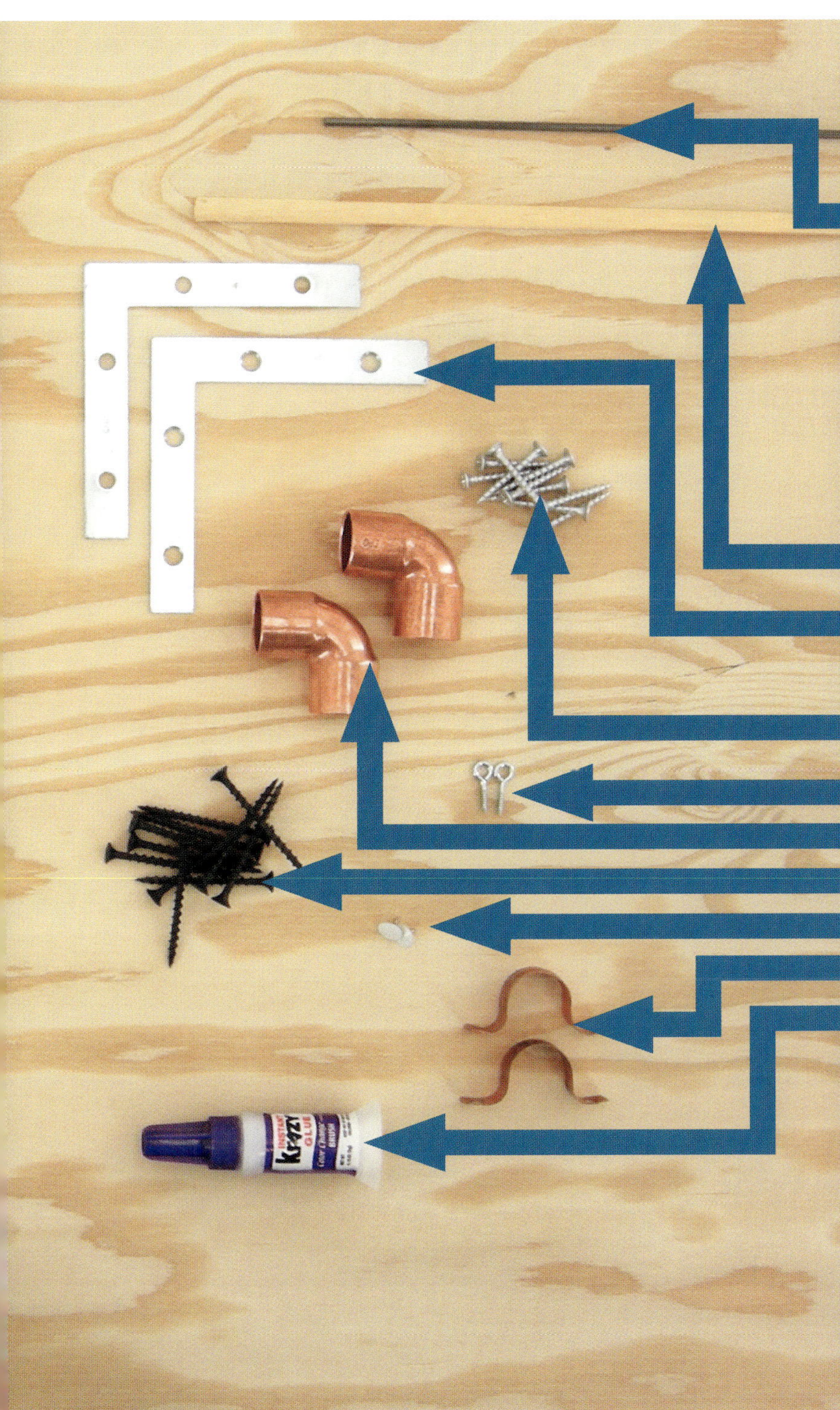

Wood 2"×4", 36" long
Piece C

Wood 2"×4", 4" long
Piece D

Metal rod, about 20"
For the crankshaft. I used a .14"-diameter iron rod, 19½" long. Other diameters may work as well, depending on ductility and strength. Metal rods come in different tempers, some more springy and more difficult to bend. Select one that bends easily, yet is strong enough to support the flywheels without excessive bowing.

25"-long, ⅜"-diameter hardwood dowels (2)

4" steel flat corner braces (2) with screws
Such as Stanley Hardware #306560

1¼" drywall screws (10)

#214 metal screw eyes (2)

¾" copper elbows (2)

2" drywall screws (8)

Thumbtacks (2)

¾" pipe clamps (2)

Cyanoacrylate glue and accelerator spray
Available in hobby stores or online

TOOLS

Hacksaw
Vise, vise-grips, needlenose pliers for rod-bending
Utility knife
Screwdriver
Drill and bits
Ruler and tape measure
Propane torch
Sandpaper
Allen wrench to fit sheave setscrew

MAKE IT.

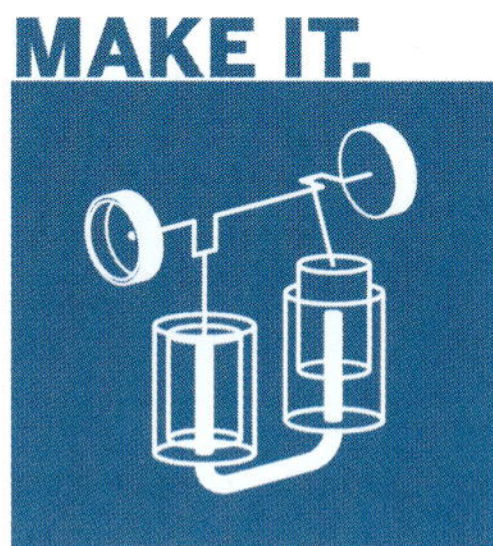

BUILD YOUR OWN STIRLING ENGINE

START Time: A Day Complexity: Easy

1. MAKE THE PISTON SUBASSEMBLIES

There are two pistons in this engine, one for the hot side and one for the cold side.

1a. With a hacksaw, carefully remove the top end of each soda can. Cut the can at the point where the flat side of the can curves to meet the top, resulting in a 4"-long piston. Sand the cut edge to remove burrs, then wash and dry the interior.

1b. Locate the center of the can bottom as accurately as possible. Push the thumbtack through the can bottom at that point. Remove the thumbtack.

1c. From the interior of the can, re-insert the thumbtack through the hole you just made.

It helps to stuff a rag into the can when pushing the tack through. This will stabilize the sides of the can and prevent buckling.

1d. Locate the center on the end of the ⅜"-diameter dowel and push the thumbtack into the wood. Carefully remove the thumbtack and coat the bottom of the dowel and the tack with super glue. Press into place and apply the super glue accelerator spray to hold fast.

1e. Test the can for watertightness. If it leaks, apply more glue.

Photography by Ty Nowotny and Jake McKenzie

1f. Locate the center of the opposite end of the ⅜"-diameter dowel, and drill a pilot hole and screw the #214 screw eye into the center. Apply super glue and accelerator spray.

2. FABRICATE THE CRANKSHAFT

The crankshaft consists of a metal rod bent in a precise way that holds the piston connecting rods in alignment.

2a. Lay out bend lines on the rod as accurately as possible using a permanent marker, as shown on the bend diagram.

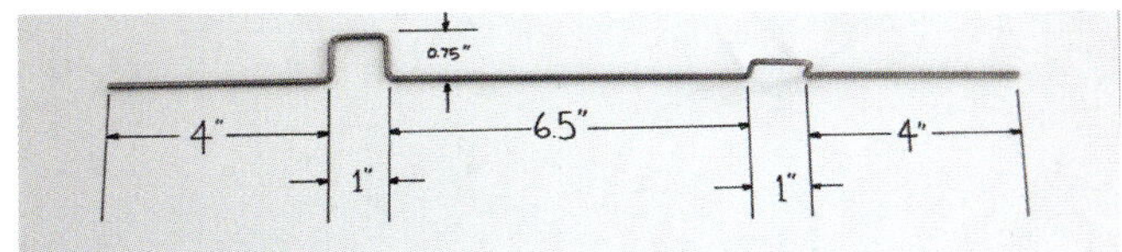

2b. Using a hammer, vise-grips, and vise, bend the metal rod as shown. Use special care when bending the rod to make the bend sizes and shapes correspond closely to the diagram. The 2 bends (the cranks) must be offset by exactly 90 degrees, and the distance from the end of the crank to the centerline of the crankshaft must be ¾".

2c. Insert the plastic spacers into the sheaves. Tighten the setscrew inside the collar of the sheave to lock the plastic spacer in place. Do not put flywheels on the crankshaft yet.

3. ASSEMBLE THE AIR CYLINDER

3a. Before soldering or gluing, cut down the 2¾" pipes if necessary, so that the overall distance of the *finished* assembly will be 7½", center-to-center.

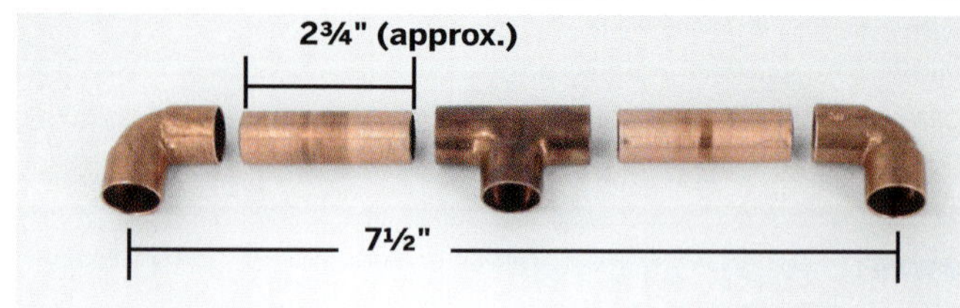

3b. Solder or epoxy the copper pipes and fittings together as shown, making certain the connections are airtight and leak-free. Note the alignment: the copper tee is rotated 90 degrees from the plane formed by the other 2 holes in the assembly.

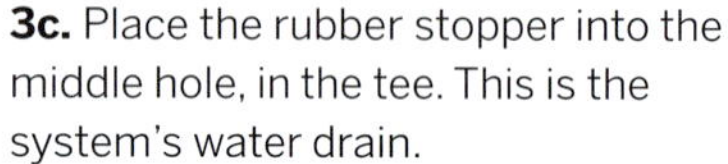

3c. Place the rubber stopper into the middle hole, in the tee. This is the system's water drain.

4. ASSEMBLE THE WATER RESERVOIRS

4a. Remove the top from each steel can, leaving the bottom intact. Sand edges smooth.

4b. Mark a ¾"-diameter circle in the center of the bottom of each can.

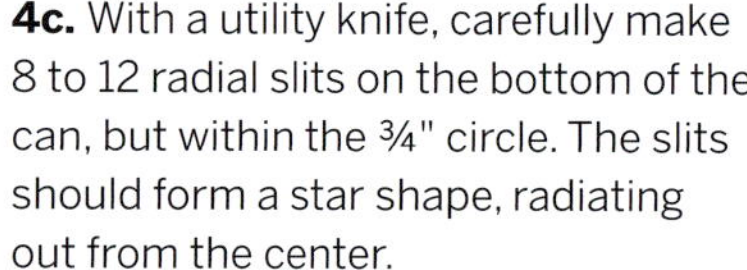

4c. With a utility knife, carefully make 8 to 12 radial slits on the bottom of the can, but within the ¾" circle. The slits should form a star shape, radiating out from the center.

If you are soldering the pipe into the can, the bottom of the can should be very heavily scored with a file to provide a toothy surface that the solder can stick to.

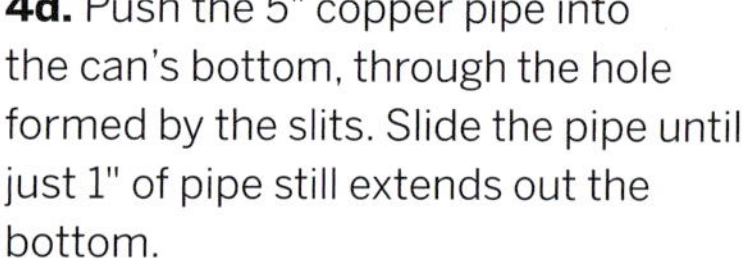

4d. Push the 5" copper pipe into the can's bottom, through the hole formed by the slits. Slide the pipe until just 1" of pipe still extends out the bottom.

4e. With the pipe concentric and parallel to the sides of the can, solder the pipe in place. (Alternatively, you can seal the pipe-to-can connection with slow-curing, waterproof epoxy glue, taking care to seal the pipe carefully so it will not leak. Allow to dry completely.)

Do not give up hope on the soldering. It is very difficult to do, but perseverance will pay off.

5. MAKE THE FRAME

5a. Using deck screws or nails, assemble wooden pieces A-D to form a frame, as shown.

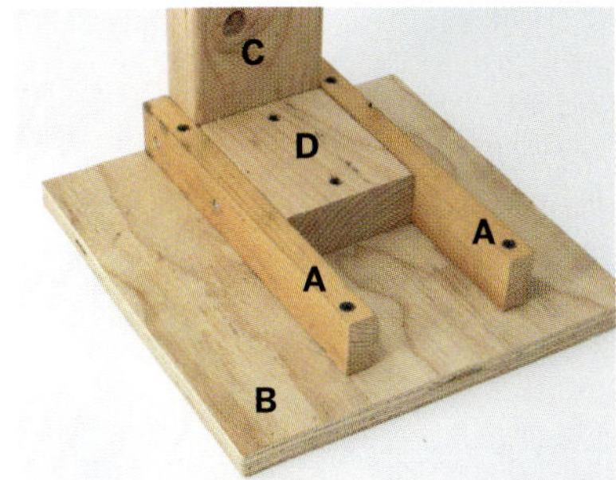

6. ASSEMBLE THE STIRLING ENGINE

6a. Insert the water reservoir assemblies into the air cylinder assembly. Fill the reservoir cans with water and check for leaks. Repair leaks with epoxy and let dry.

6b. Measure and then mark a spot on each 1"×2" frame piece, 3¾" from the back edge of the frame. Place the combined water reservoir and air cylinder assembly on the 1"×2" frame pieces at the marked spots. Now place the ¾" copper pipe clamps over the assembly. Screw the pipe clamps into the 1"×2" pieces. The clamps must hold the combined assembly firmly in place.

6c. Slide the screw eyes on the connecting rods onto the crankshaft, so that 1 screw eye is on each of the 2 cranks. Place the soda-can pistons inside each of the water reservoirs so that each soda can rests on copper pipes. Turn the crankshaft so that one of the cranks is pointing downward.

Holding the crankshaft level, lift the crankshaft until the can corresponding to the bottomed crank is about ½" above the top of the copper pipe. This is the desired height for the crankshaft. Mark this height on the upright 2"×4" and attach the angle bracket at this point, making sure that the hole through which the crankshaft will pass is located 3¾" from the back of the 2"×4".

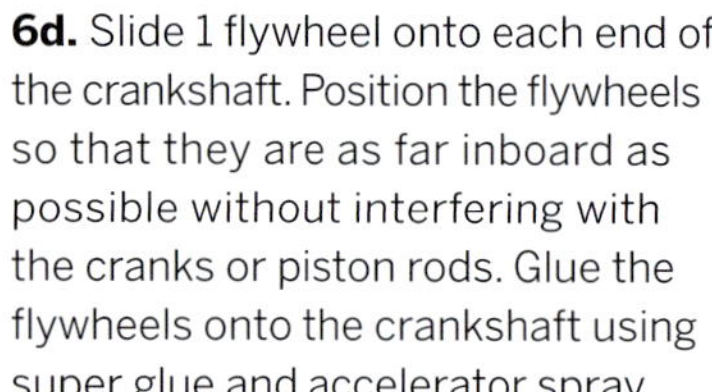

6d. Slide 1 flywheel onto each end of the crankshaft. Position the flywheels so that they are as far inboard as possible without interfering with the cranks or piston rods. Glue the flywheels onto the crankshaft using super glue and accelerator spray.

You're done!

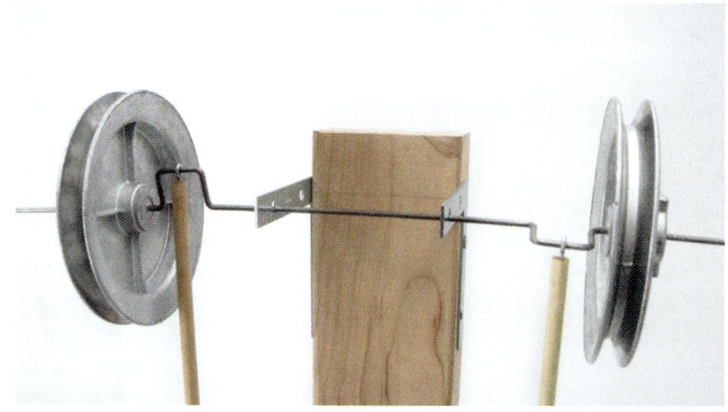

FINISH X

NOW GO USE IT »

USE IT.

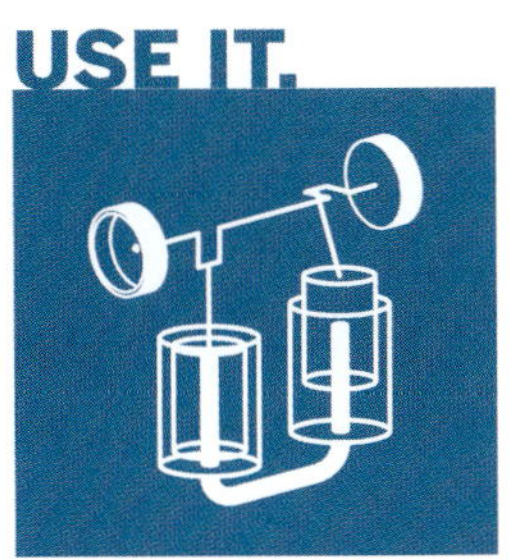

MAKER, START YOUR ENGINE

To start your Stirling engine, turn the crankshaft until both cranks are tilted upwards at 45-degree angles to the vertical. With the stopper removed from the drain, fill each side with water, until a trickle runs out the drain. Dry it and replace the stopper.

Designate one side as the hot side, then heat the water on that side to boiling with a propane torch. This takes a while, depending on the heat output of the torch. Be patient.

When the water is ready, start the engine by giving the flywheels a small push. The rotation is determined by this rule: the cold side is 90 degrees behind the hot side.

If built properly, your engine will dip and lift, dip and lift, 20 to 30 times per minute to the chuff-chuff beat of Robert Stirling's ancient idea.

TROUBLESHOOTING

1. Make sure the engine is level. The crankshaft must revolve freely, and the connecting rods should stay in the middle of each crank as it rotates. Use shims or cardboard to level the system. If the connecting rods will not stay centered on the cranks, you can add a small wire loop or small nut to the rod on either side of the eye screw, fastening them into place with super glue.

2. You may have to experiment to find the best flywheel weight. If the flywheels are too heavy, the metal rod will bow, interfering with the crankshaft's rotation. But if the flywheels are too light, there won't be enough inertia to carry the crankshaft past the volume compression phase and into the next expansion stroke. If this happens, the engine will pulse but not run cyclically. You can add weight to the flywheel by simply taping bolts or other weighty objects to its perimeter.

3. Large steel cans full of water take time to heat. Be patient, and let the water heat to 200°F or more.

4. Minimize friction and interference. Friction is your engine's greatest enemy. Minimize rubbing between pistons and water cans, between connecting rods and cranks, and between the crankshaft and the metal support angles that attach it to the wooden frame.

5. Add a regenerator. A regenerator consists of a small piece of heat-conducting metal gauze placed in the air cylinder just behind the rubber stopper. A regenerator will improve cycle efficiency and make the machine turn faster. The copper gauze sold for cleaning kitchen pots ("Chore Boy") works well.

HOME MYCOLOGY LAB

By Philip Ross

Photography by Kirk von Rohr

CULTURAL REVOLUTION

Use an off-the-shelf home air purifier to make a laminar flow hood for your own miniature mycology lab. Then use it to culture and grow mushrooms, and to perform other experiments that require a clean-room environment.

Agriculturalists have long considered mushroom growing a challenge, largely because you need a space that's as hygienic as a hospital or a chip-fab clean room. Laboratories create these spaces with a piece of benchtop equipment known as a laminar flow hood, but these are prohibitively expensive. This article explains how you can make your own "hood" out of a household air purifier and use basic kitchen techniques to culture and grow mushrooms.

The crucial component in the purifier is its High Efficiency Particulate Air (HEPA) filter. Originally developed for the Manhattan Project during World War II, these filters later became standard in hospital and computer manufacturing facilities. Now, cheap HEPA filters are built into vacuum cleaners and other mass-market consumer appliances, and their trickle-down availability lets amateur biologists run procedures that were previously reserved for large corporations, universities, and research institutions.

Set up: p.105 **Make it:** p.106 **Use it:** p.110

Philip Ross is a San Francisco Bay Area artist who makes sculptural artifacts using living organic materials, including fungus, shellfish, plants, and compost.

SPORES IN THE HOOD

A home mycology lab might not make you a fun guy, but it *can* help you grow your own mushrooms.

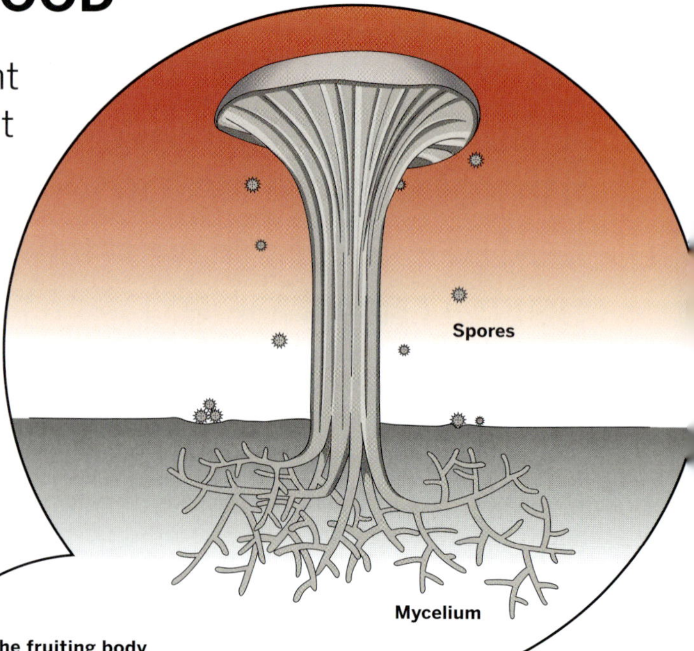

Fruiting Bodies

Mushrooms are the reproductive organs (fruiting bodies) of a much larger organism — a network of cells, called the mycelium, that lives underground or inside dead and dying trees and digests cellulose and other plant matter. When it rains, the mycelium starts to grow a fruiting body. This is why mushrooms seem to grow so fast.

The fruiting body creates and disperses millions of spores. The wind carries them to new locations, where some will germinate and grow into new mycelium. Unlike plant seeds, spores are microscopic and nearly invisible, so it's no surprise that human agriculture developed with plants rather than fungi.

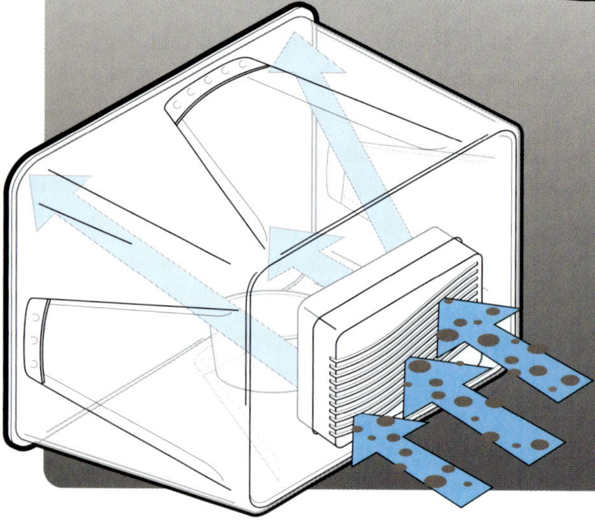

A Fortress of Clean

It seems strange that we need a clean space to grow things that emerge from dirt or dead tree flesh. But countless microorganisms and spores fill the world around us, and only a sterile environment allows us to grow certain organisms in a controlled, reproducible manner.

Our clean box acts as a fortress — inside the enclosed environment, purified air is drawn in through a HEPA filter on one side, which traps the living microorganisms, dirt, and spores floating around.

The purifier's fan maintains positive airflow out of all the box's other openings, which gives airborne germs no way of getting in.

Illustration by Tim Lillis

SET UP.

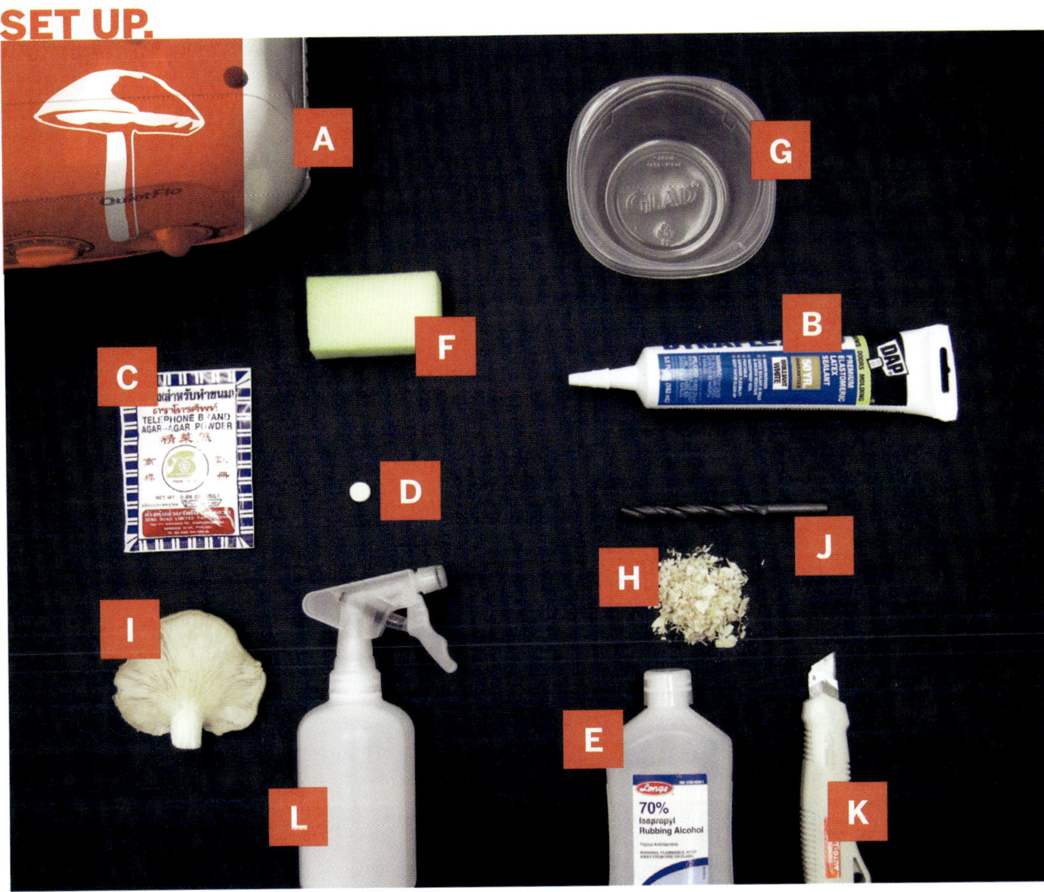

MATERIALS

[A] Small countertop home air purifier with HEPA filter and flat, boxy shape, such as the Sunpentown AC-3000 Magic Clean Available at home, hardware, and large drugstores

[B] White silicone caulk/ sealant

[C] Agar (seaweed gelatin) Available at many health food stores and Asian groceries

[D] Tums tablet, or other form of calcium carbonate

[E] Isopropyl alcohol

[F] Cotton balls or cut-up sponge pieces

[G] Small, transparent dishwasher-safe plastic container with lid (one or more) Like the kind fresh salsa comes in, or use lab-standard petri dishes

[H] Cellulose-based growing substrate such as sawdust You can also use paper- or pine-based cat litter, wood shavings, newspaper, rye grain, or birdseed; look in pet supply and animal feed stores. Some animal bedding is heat-treated to kill microorganisms, which is a plus. If you plan to eat the mushrooms, make sure that there is nothing in the substrate that you wouldn't want to ingest yourself.

[I] A fresh mushroom I recommend oyster mushroom for first-timers, available at upscale food markets, natural food stores, Asian groceries, farmers markets, or in the wild. It should smell fresh, like a forest. A fishy or sour smell means the mushroom probably has too much bacteria for good culturing.

Translucent plastic file box with lid Bottom must be bigger than output side of air purifier; available at office, home, or hardware stores

Aluminum foil

Warm, soapy water

A bleach-and-water solution

TOOLS

[J] Drill and drill bits

[K] Hobby knife

[L] Spray bottle with soft (low mineral content) water If you are uncertain about the "hardness" of your water, use filtered, deionized, or spring water.

Latex gloves

Pen

Stove and stockpot with steamer basket, or steamer

1-quart Mason jar with lid

Labels and journal

Keyhole saw or jigsaw

MAKE IT.

BUILD AND USE YOUR CLEAN BOX

START **Time:** 1 hour to build; 2 weeks to grow **Complexity:** Easy

1. CUT THE HOLE

1a. Find the output side of the air purifier, and trace it on the bottom of the plastic box.

1b. Drill pilot holes at the corners of the traced outline.

1c. Use a keyhole saw or jigsaw at the highest speed to cut out the entire hole.

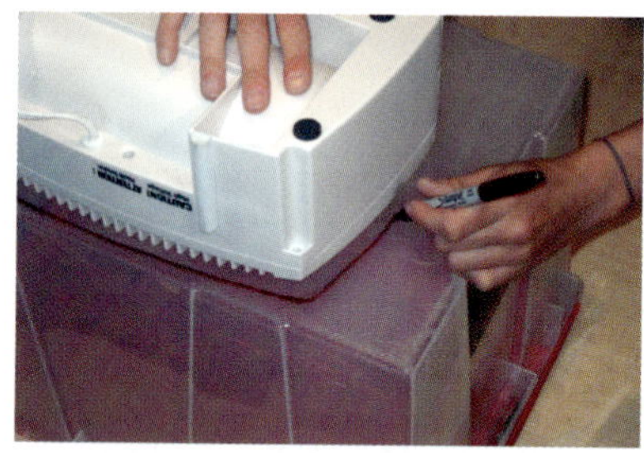

2. INSTALL THE PURIFIER

2a. Fit the air purifier into the hole, with the intake side facing out and the output side blowing into the box. You might want to prop it up on some books to keep it in place.

2b. Use silicone sealer to generously caulk around the air filter, securing it in place. Let it sit overnight so that the caulk can dry. That's it — now you have your hood! Move it onto a good work surface with its opening facing you, and let's start using it.

Photography by Philip Ross

3. CLEAN THE HOOD

This isn't just Step 3; it's something you'll need to do every time you work inside your laminar flow hood. The hood is crucial for mushroom growing, but it's only one part of the larger regimen of cleanliness required for successful lab work.

3a. Clean all of the hood's surfaces with warm, soapy water.

3b. Disinfect all surfaces of the hood with a bleach-and-water solution.

3c. Finally, turn the fan on and disinfect the hood with isopropyl alcohol. You can never be too clean!

4. MAKE THE AGAR PLATE

We'll begin growing our mushroom tissue in agar (seaweed gelatin), a standard laboratory growth medium. Petri dishes are traditionally used, but you can use any shallow, washable container with a lid. As long as you're cooking a batch of agar, you'll find it handy to make several of these plates at once and store them in airtight bags for later use.

4a. Drill or cut a ½" hole in the lid of a washable plastic container.

4b. Wash the container and lid with soap and water, and then sterilize by immersing them in simmering water for 3 minutes. Switch on your hood's fan, and move the container and lid inside for drying.

4c. Make a filter by soaking a piece of cotton or sponge in isopropyl alcohol and then wringing it out. Place the filter in the hole in the container lid. It should fit snugly.

The sponge-piece filter keeps the mushroom tissue protected while letting it exchange gases with the surrounding air.

4d. Mix 1 tablespoon of agar in 1 cup of water. Bring to a low boil and slowly simmer for about 15 minutes, stirring occasionally. Add a large pinch of the growing substrate you'll be using later (sawdust, cat litter, barley, etc.) to the simmering agar as a source of nutrition.

4e. Inside your hood, pour the hot agar into the newly sterilized container until it is about as thick as a pencil. Let the gelatin cool and congeal.

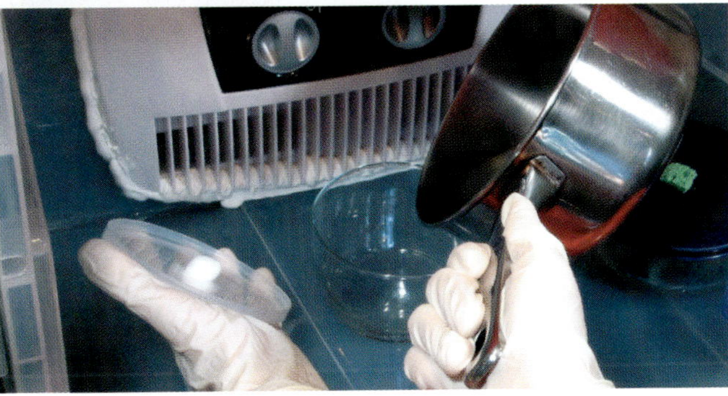

5. START THE CULTURE

5a. Sterilize a hobby knife by soaking it in isopropyl alcohol for a few minutes and letting it dry in the hood.

5b. Inside the hood, break open the stem of a mushroom. Cut a clean, unbruised chunk of tissue from inside the stem and place it onto the agar in one of your plates. It is important to use only tissue that has never been exposed to air before.

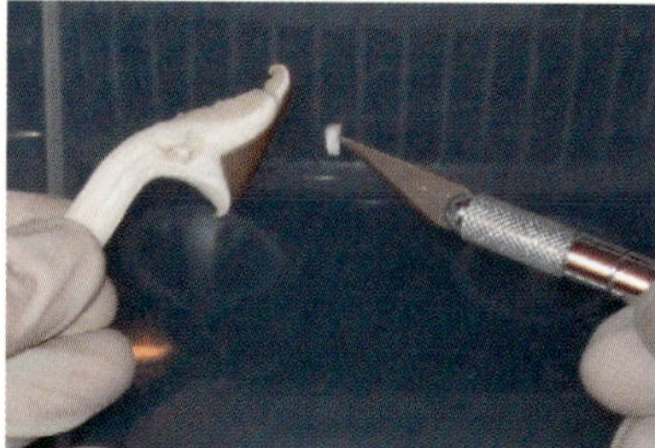

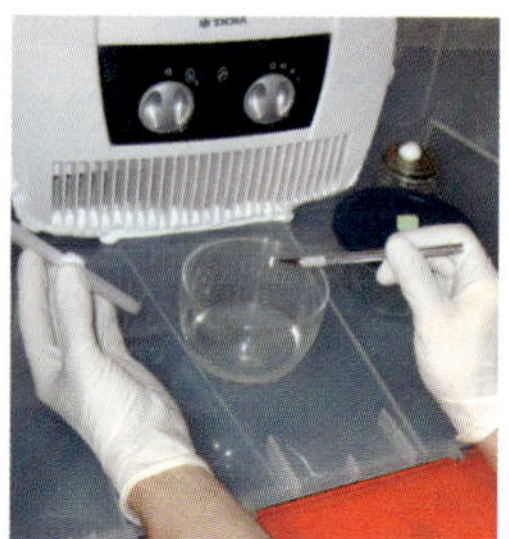

5c. Cover the container with the filter lid and leave it in the hood. Cells from the mushroom tissue will grow out across the agar and will look like a thready mold after a few days. When these growing cells have reached the sides of the container, you're ready for the next step: moving them onto a larger food source.

6. TRANSFER TO THE GROWTH MEDIUM

6a. Drill a ½" hole in the lid of the Mason jar, and make and fit a filter for it, as in Step 4c.

6b. Add 1 cup of growth medium into the jar, along with ⅔ to ¾ cup water, and ⅛ teaspoon crushed Tums or other form of calcium carbonate. Mix the ingredients, seal the jar, and tightly wrap the lid with foil.

6c. Stand the jar upright in the pot and steam for 1½ hours, covered. All the water in the jar should be absorbed into the substrate. If there is any standing water at the bottom of the jar, pour it out and steam for another 15 minutes. Pooled water increases the chance of bacteria growing in the jar.

The calcium dissolves and helps keep the pH of the mix neutral; the foil keeps excess water from getting in during the next step, sterilization.

6d. Immediately after steaming, move the jar into your hood with the air stream on. Let it cool (1 to 1½ hours), then shake the jar to loosen the substrate material. Remove the foil.

6e. Sterilize a knife as in Step 5a, and use it to cut a wedge out of your agar plate, working inside the hood. Transfer the slice into the jar so that it sits on top of the substrate.

6f. Reseal the jar and move it out of your hood to a dedicated growing area. I like to use another well-cleaned, clear plastic box for this purpose because it adds an extra layer of protection. The growing area should have a comfortable temperature range (60-80°F) and a day/night light cycle, but direct sunlight is not so good. If there is no ambient daylight in your growing area, then a light with an on-off timer will do. Every 3 days it helps to gently shake the jar, mixing the growing mycelium through the substrate.

6g. After about 6 days to 2 weeks, the mycelium will have fully grown through the material in the jar. Now you have a choice: use this material as spawn to grow more mushroom material, or encourage the fruiting bodies to form. If you use it as spawn, divide the material in the jar into 4 more jars of sterilized substrate, as prepared in Steps 6a-6d.

6h. To encourage mushrooms/fruiting bodies to grow, expose the mycelium to more air, either by removing the air filter or by unsealing the lid and letting it sit askew on the jar's rim. You will also need to water your mushrooms daily using a spray bottle. Don't get too much water into the jars or spray too close or too hard. The mushrooms might start appearing after a few days or a few weeks, depending on various factors, so the best thing to do is be patient. Open the box they are growing in only when you are watering them, and resist the urge to touch any of the living material until you are ready to harvest the mushrooms.

6i. Another way to encourage mushroom growth is to temporarily change the temperature. Place the sealed jar in the fridge overnight (or outside if it's cold), and protect it from cooties by putting it in a sealed plastic bag. The next morning, remove the jar from the bag and put it back into its grow box.

FINISH X

NOW GO EXPERIMENT »

When I Hear the Word Culture ...

A growing number of artists use living organisms and the life sciences to make their work. Many want to call attention to world-altering trends in biotech that permit living processes to be controlled and commodified by industry. The public knows about the intellectual property issues surrounding digital media, but now that it is possible to copyright living entities, normal growth and reproduction (of protected organisms) are also crimes. The inevitable fights over biopiracy will make the Napster flap seem trivial.

Things are already starting to boil. The Critical Arts Ensemble (CAE) is a collective of bioartists who, like many scientists and educators, use common laboratory equipment and exchange biological materials through the mail. Nevertheless, there is an ongoing federal grand jury case against CAE member Steve Kurtz, initiated on charges of terrorist activities and now focusing on mail and wire fraud. Scientists and artists worldwide have come to Kurtz' defense, because a guilty verdict would have a chilling effect on both creative expression and scientific research.

Many plants, including orchids, can be cloned like mushrooms, and recent years have also seen advances in techniques for culturing human tissue. Skin and cartilage are grown routinely for reconstructive surgeries, and researchers are working on growing entire organs. At the University of Western Australia, a collective of artists and scientists called SymbioticA uses human tissue as a creative medium, inviting us all to ponder the complexity of what it means to be cultured.

Many of my own artworks are grown into being; see philross.org.

EXPLORATORY GROWTH

EQUIPMENT ALTERNATIVES

The techniques described here are a good place to begin, and can be altered to fit your budget and space. Companies such as Fungi Perfecti (fungiperfecti.com) and Carolina Biological Supply (carolina.com) sell prepoured agar plates, mycelium spawn, and plastic bags with preattached filters, which are common in the mushroom industry — or you can improvise your own with zip-lock bags and filter material. Drugstores have all sorts of supplies for protecting living materials. Experiment, keep track of what you've done, and be safe.

LABELING AND JOURNAL KEEPING

As you experiment, keep track of what you are doing. Write the date and type of mushroom you are culturing onto your agar plates and jars. In a journal, write down the recipes you followed or changed, the equipment and techniques you tried, measurements of ingredients, cleaning techniques, the number of times you cooked things, the smells of your cultures, the substrates you used, how things grew or got contaminated, and any other things you notice — even the weather and temperature outside. Without this documentation, you'll lose track of what's what. And when something works well, you can refer to your log to try to reproduce the results.

MUSHROOM GROWING TIPS

Clean the room and all of your equipment so that you could literally eat off of any surface.

» If there are cobwebs in the room, there are probably microorganism-carrying spiders and insects. Get rid of them or try another location.

» Choose a work area that's isolated from open windows, drafts, plants, pets, and other people.

» Use dedicated equipment — things you're not also using for cooking or other activities.

» Work on a smooth surface that can be bleached down. For a good smooth surface, you can tape down plastic sheeting or an opened garbage bag.

» Follow strict personal hygiene before doing any lab work: shower, brush your teeth, pull back long hair, and clip your nails. When washing your hands, scrub up to your elbows and rake your fingertips across a bar of soap to clean under the nails. After drying off, follow up with isopropyl alcohol. (Once you are familiar with culturing techniques, you can be a bit less orthodox about cleanliness.)

» Don't touch anything unnecessary: your face, phone, doorknobs. Remove watches and jewelry.

» To supply nutrients, try adding a pinch of 20-20-20 plant food or crushed multivitamin into your cellulose, or put a piece of dry cat food into your agar.

» If any of your agar or mushrooms are contaminated with mold, discard them immediately. These will infect the other things you are trying to grow.

» Join a local mycological society. These are great places to learn about mushroom growing, as well as wild mushroom identification.

» Only eat mushrooms that you can identify confidently.

RESOURCES

The Science of Life, by Frank G. Bottone, Jr. — Biological principles and good projects for kids.

Mycelium Running and other books by Paul Stamets — Essential reading by the godfather of mushroom cultivation.

Critical Art Ensemble — critical-art.net, caedefensefund.org

SymbioticA — www.symbiotica.uwa.edu.au

Science and technology artist directory (including bio-related) — userwww.sfsu.edu/~infoarts/links/wilson.artlinks2.html

Grain substrate preparation using a pressure cooker — kalyx.com/catalog/grain.htm

The Pontani Sisters use multiple weapons in their super-secret after-hours life of crimefighting.

HEAD-MOUNTED WATER CANNON

Use steel fire-extinguisher power to pummel plastic squirt toys. By John Young

Photograph by Julie Gottesman

Let's face it: At some point this summer, you're going to be in a water fight. Whether it's at a family barbecue or an office picnic, some 12-year-old is going to leer at you from behind 25 bucks worth of store-bought plastic, and that little punk is going to think that the orange and blue Mega Awesome Hydrolator 9000 they're clutching is the last word, the *ultima ratio regnum*, in neighborhood water warfare.

Think again, punk. With about two hours of effort and the parts listed on the next page, you can hack together a water weapon of such power, such style, such extraordinary and exuberant overkill, that you'll be out of the store-bought leagues forever. Lock yourself in the garage, play the *A-Team* theme, and emerge at the end of your build montage with a pressurized, stainless steel, head-mounted water cannon that packs five gallons of icy-cold water at 100 psi.

Sourcing the Main Components

The big pieces in this project are the extinguisher, the backplate, and the helmet. For the fire extinguisher, you're looking for a standard, stainless steel water extinguisher with a wide collar. Water extinguishers aren't legal for fighting fires anymore, but many fire safety companies have a dozen of these old models in the back room. So

MATERIALS

- Fire extinguisher Water type with wide collar
- Plastic scuba backplate These can run $150 new, but most dive shops have battered old ones that they'll let you have for a few dollars, or for free.
- Motorcycle helmet
- Bicycle brake lever assembly, brake cable, and cable housing
- Wire nut, small
- Wooden rod Bicycle tubing width, cut from a plunger, broomstick, or similar
- 5"×5"×1½" wood block
- Jigsaw Or other way of cutting wood to fit curve of helmet
- Wood screws
- Epoxy (optional)
- Angle brackets and pop rivets (optional)
- ⅝" garden hose repair fittings, male (5)
- Quick-coupler (aka quick-connect) sets for standard ⅝" garden hose (2)
- Standard garden hose nozzle
- 4' length of ½" vinyl tubing Should be strong but stretchy enough to fit tight over the hose fittings
- Duct tape
- Drill and drill bits
- Scissors, pencil, index card
- Bicycle pump or air compressor
- Vacuum grease From an HVAC supply store (or your building's facilities office)

the best way to find one is just open your phone book to "Fire Safety" and start calling. Be sure to explain that you don't want it to fight fires, but to fight injustice (or to cool welds, which is what most folks buy these old extinguishers for). A water extinguisher by itself puts you in the Big Leagues of water combat, and you can recharge it with a bicycle pump or air compressor. Haul one out, and you've already won the water-fight arms race; the rest is all about style.

The wider the collar on the extinguisher, the easier it will be to recharge; avoid models with narrow necks that require a pipe wrench to remove. A working air gauge is also a plus. Avoid antiques; you're looking for something in good, serviceable shape.

The scuba backplate will clamp directly onto the extinguisher, and hold it on your back, making you look like a cross between a firefighter and some kind of vigilante astronaut. Get an old backplate at a dive shop. Note also — and this is very important — that making friends at a dive shop will give you access to a whole new category of deeply awesome spare parts for noodling around with. Speargun power bands! Waterproof thruster handles! Dangerous neoprene glue! Every project is cooler with scuba parts.

Any motorcycle helmet will do, but to really nail the Evel Knievel look, search eBay or your local Elks lodge for a '70s Buco or Bell metalflake helmet with a bubble visor. You want something that looks like it could have been used to jump the Snake River Canyon. With any luck, your neighbor will have a vintage stars-and-stripes Shoei sitting on top of some dusty water skis above the garage. You can offer a trade with a new, cheap helmet — or just offer to use your new hydro-offensive powers to keep their lawn free of dogs for a while.

Assembly

1. Attach the bicycle brake lever to the wooden rod, and install the cable in the lever. Drill a 3/16" hole through the ends of both handles of the fire extinguisher so that the two holes line up. Clip off the other end of the brake cable, thread it through the housing, and then thread it up and out through the extinguisher handles, with the housing tight against the inner lever. Trim the cable so it just emerges from the outer handle, crimp a wire nut onto it, and cover it with duct tape. When you're done, you'll be able to squeeze the extinguisher's handle by working the brake lever.

2. Drill a ⅝" hole the long way through the wooden block, parallel to one side. The hose will run through this hole. Cut the index card until it matches the curve of the helmet, trimming away until the fit is right. Transfer the curve to the block with a pencil, along the side opposite the hole, and cut the block to fit. (Using a cardboard template without measuring is known as "the Jesse James method.")

3. Depending on how handy you are with wood tools, you can either carve the block to match the lateral curve of the helmet, or you can simply caulk in some two-part hardening epoxy, like PC-7, to mate the block of wood to the helmet. Finish by either screwing angle brackets onto the block, then pop-riveting them onto the helmet, or running wood screws from the inside of the helmet up into the block. Either way, the mount should be strong: eventually, everyone gets tackled by sore losers.

4. Cut the extinguisher's rubber tube a few inches from where it leaves the extinguisher and add one repair fitting to the stub. Attach one half of a

Fig. A: Bicycle brake cable snakes up from brake-lever trigger and threads through the extinguisher handles. Fig. B: Cable guide keeps the vinyl tubing in place at the base of the helmet.

Fig C: Extinguisher, extender, backplate, and trigger rod are assembled and ready to go. Screw the garden hose quick-connect onto the helmet tube, and you're ready for combat.

hose quick-coupler to that fitting.

5. Cut the vinyl tubing into two 2' pieces. On one piece, add repair fittings and quick-couplers to both ends; that's your extender, which will run from the tank to the helmet. On the other piece, add a repair fitting to one end, connect it to the hose nozzle, and run the other end through the wooden block on your motorcycle helmet, from front to back. You'll have to really jam it in there, and friction will probably hold it in place, or else you can always use epoxy.

6. Attach the last repair fitting to the tube running through the back of the helmet, put on the last quick-disconnect, and your helmet is ready to go. You might want to stabilize the tube by running it through a cable guide at the base of the helmet.

(Optional) Epoxy a scrub brush to the top if you want a Roman centurion look. Or clip on some Pelican flashlights if you're going to battle after dark.

Now Go Have Some Adventures

Fill the extinguisher to the index mark on the inside, spin on the collar (some vacuum grease will help you keep a tight seal), and pressurize the tank to 100 psi. Snap the extender to the tank and the helmet, attach the backplate, strap it onto your back, don the helmet, and you're ready to rock and roll. You can rule your block, sell justice to the highest bidder, or loan your rig out to those supplicants whose cause is worthy.

Just try to keep your eyes on your opponent when you pull the cable release: the water pressure tends to snap your head back a bit! You probably couldn't hurt someone with a store-bought water gun (unless you clubbed them with it), but you could definitely hurt someone with your water cannon if you shot them in the eyes at close range. Or if you ask them to carry it for you — it's *heavy*. Be careful out there!

+ The author lends his head-mounted water cannon on a two-week basis to those whose causes he deems worthy. Read more at ultimatewatergun.com.

John Young is a technologist at Digitas in New York. He enjoys web development, motorcycles, and heavy metal music about European history.

Ruggedized access point and antenna cast wide network from literally eaves-dropping location.

WEATHERPROOF WI-FI ACCESS POINT

Outdoor router with minimal coaxial run maximizes network range. By Will O'Brien

In typical neighborhood wireless network deployments, the routers/transceivers operate indoors and connect to outdoor antennas via coaxial cable. The problem with this arrangement is that a lot of power gets lost in the cable, reducing the range. The combination of high frequencies (wi-fi operates at 2.4GHz) and low power (imposed by the FCC) makes coaxial cable very inefficient. For example, the maximum output of a Linksys WRT54GS router is 84 milliwatts, but after you add 20 feet of Times Microwave LMR-400 cable, you reduce its broadcast power to 62mW. Standard RG-6A cable is even worse (see Resources for the calculators used to derive these figures).

To solve the loss problem, you can either add an expensive, FCC-approved amplifier, or you can just eliminate the long coaxial connection. I took the cheaper, more flexible, latter approach. I located a full access point outside, right next to the antenna, inside a weatherproof enclosure.

Choosing the Main Components

For the router, I chose the hacker-favorite Linksys WRT54GS, which is upgradable with third-party firmware such as Sveasoft (sveasoft.com), Hyper-WRT (hyperwrt.org), and OpenWrt (openwrt.org).

For the enclosure, I found a great NEMA 6 aluminum pole-mount enclosure from FAB Corp

Photography by Will O'Brien

MATERIALS

Linksys WRT54GS or other wi-fi router I ordered mine from amazon.com for $59
NEMA 6 aluminum enclosure FAB Corp (fab-corp.com), $20
Outdoor 2.4GHz antenna FAB Corp, $50 and up
Bulkhead N Female to N Female connector FAB Corp, $5
RP-TNC Male to N Male adapter FAB Corp, $5
Nylon PG-7 or PG-9 cable gland (gasket) $1 from Allied Electronics (alliedelec.com) or $3 from Metrix Communication (metrix.net)
Power over Ethernet (PoE) adapters I made my own, following NYCwireless' excellent how-to (see Resources); alternatively, you can buy the Linksys Power Over Ethernet Adapter Kit, linksys.com, $35
Plastic cutting board About a $1 from discount stores
Bulk Cat-5E Ethernet cable Enough to run from network connection to router; mine is from Home Depot
2" 6-32 machine screws and nuts (3)
Phillips screwdriver
Pliers
Hammer
Wrench socket (½") or center punch
Saw
Drill and drill bits
Dremel tool with cut-off wheel
RJ-45 crimper
Pencil, paper, and cardboard
Scissors

(fab-corp.com) for a paltry $20. The NEMA 6 rating, from the National Electrical Manufacturers Association, means that the enclosure can resist direct dousing from a hose and can even handle being temporarily submerged. It's just about perfect for a ruggedized WRT54G or WRT54GS.

I already had a small omnidirectional 2.4GHz antenna with a female N connector, so I decided to use that. To hook it up to the access point (AP), I needed a few inches of coax, so I used a short male-to-male N adapter made out of some LMR-400 fitted with male N connectors at each end.

If you don't already have your antenna, a pigtail type would be cheaper. With one of these, you'd seal around the hole the pigtail runs through. This would attach the antenna semipermanently to the box, making it less easily swappable.

The WRT54GS router needs two things to function: 12 volts DC and a network connection. Conveniently, standard Ethernet cable has spare wire pairs that can carry the power, putting everything the router needs into one cable. The standard for this concept is called Power over Ethernet (PoE).

Assembly

1. Make a cardboard template of the inside of your case, and use it to saw the plastic cutting board down so that it just fits in the bottom.

2. Make a paper template for the case's mounting holes, and use it to mark the same positions on the cutting board. Drill and countersink these locations to fit the screws that came with the case. Test-fit the mounting holes with the screws.

3. If your router is new, test it out. We're about to take it apart and void its warranty.

4. Unscrew the router's 2 antennas. Use pliers to squeeze the plastic collars if necessary.

5. Newer Linksys units have 2 screws under 2 of the feet. Remove them, then hold the back half of the unit and push forward on the front feet with your thumbs. It should pop apart.

6. Take the bare Linksys board and test-fit it inside the case. See how the antenna connectors stick out and block it from fitting? You could just drill holes for the stock antennas, seal them with silicone glue, and be ready to go. But we want a solid N connector for a larger antenna.

7. Now you're *really* going to void the warranty. We need to remove the antenna connectors, but they have too much metal to desolder. Instead, use the Dremel tool and cut-off wheel to carefully cut the corners that they're attached to off of the main board. One of the connectors gets its signal from the board underneath, and cutting it off severs the connection. We don't need that one. The other one connects to the board via coaxial cable. Don't cut the cable; that's the one we're using.

8. Attach the Linksys board to the cutting-board mount. The Linksys has 4 mounting holes. The one that's nearest to the shiny RF shielding, toward the front of the router on the coax cable side, has a trace running next to it that you don't want to risk shorting or damaging, so leave that hole alone. Position the router board on the cutting board, mark and drill the locations of the other 3 holes to accommodate the machine screws, and bolt the boards together.

9. Punch out the holes for the network cable and the antenna. One side of the case has hole knockouts already prescored. Find the ⅝" one and the ½" one, place a socket or center punch on each, and whack with a hammer to pop them out.

10. Mount the board in the case using the holes you drilled in step 2. Install the bulkhead Female

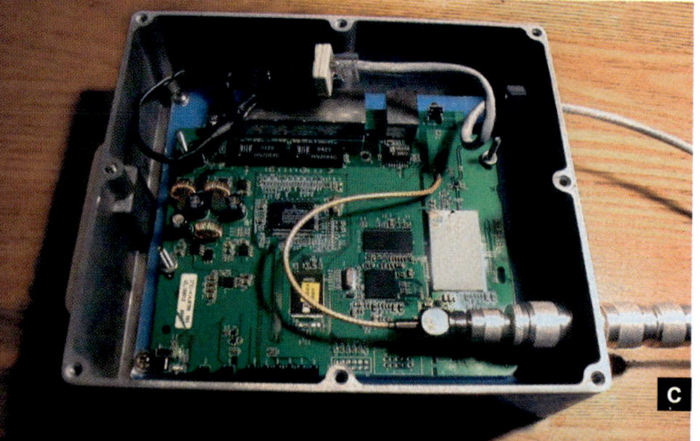

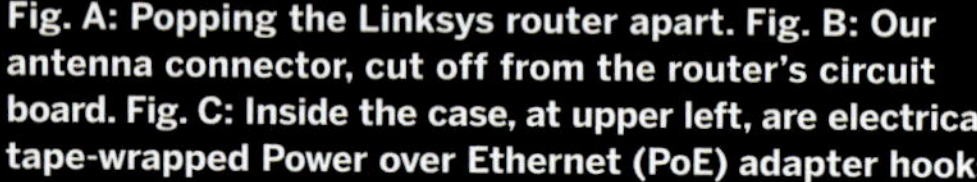

Fig. A: Popping the Linksys router apart. Fig. B: Our antenna connector, cut off from the router's circuit board. Fig. C: Inside the case, at upper left, are electrical tape-wrapped Power over Ethernet (PoE) adapter hooks up to router RJ-45 and power ports; at lower right, the board's antenna connector plugs into an N adapter and bulkhead connector which take the connect outside. Fig. D: The sealed case.

N connector in the ⅝" hole in the case. Connect the now-loose antenna cable from the Linksys board to the RP-TNC Male to N Male adapter, and plug it into the bulkhead connector from the inside.

11. Construct your 2 PoE adapters, if you're making your own (see Resources). I used a pair of RJ-45 to DB-9 adapters and cut the plastic housings in half. Then I crimped on some male RJ-45s, soldered on the power connectors, and wrapped both pieces with some quality electrical tape. When you're done, double- and triple-check the wiring with a multimeter; you don't want to fry your router. Connect the adapter with the power plug to your router's RJ-45 and power port. For my home setup, I used a LAN port because my firewall is the actual gateway.

12. Run the Ethernet/Cat-5 cable through the gland, which will seal against the hole in the case. (I had to take the gland apart and slide the rubber gasket down the cable, then strip and crimp the RJ-45 end back onto the cable.) Assemble the gland about 6"-8" from the end of the cable, threaded through the ½" hole in the case.

13. It's time for testing. Hook up your antenna to the N jack on the outside of the box. Plug one end of your Ethernet cable into the router's PoE adapter. Connect the other PoE adapter, the injector, into power and your network. Plug the other end of your Ethernet cable into the injector.

14. If it all worked, screw the lid onto the case. Crimp both ends of your Ethernet cable, if you haven't already. I had a 100-foot run of cable to my access point, and I took a risk by not crimping the injector end until after I'd already made the run. But I had already tested the box with another cable, so I trusted that it would work — which it did.

If you've done everything right, you should have a happily working WRT54GS access point inside a nice, rugged enclosure.

Resources

dB to mW, and other wireless calculators:
wisp-router.com/calculators and makezine.com/go/calculator

NYCwireless PoE (Power over Ethernet) how-to:
nycwireless.net/poe

Will O'Brien pulls espresso and modifies innocent kitchen appliances somewhere in middle Missouri.

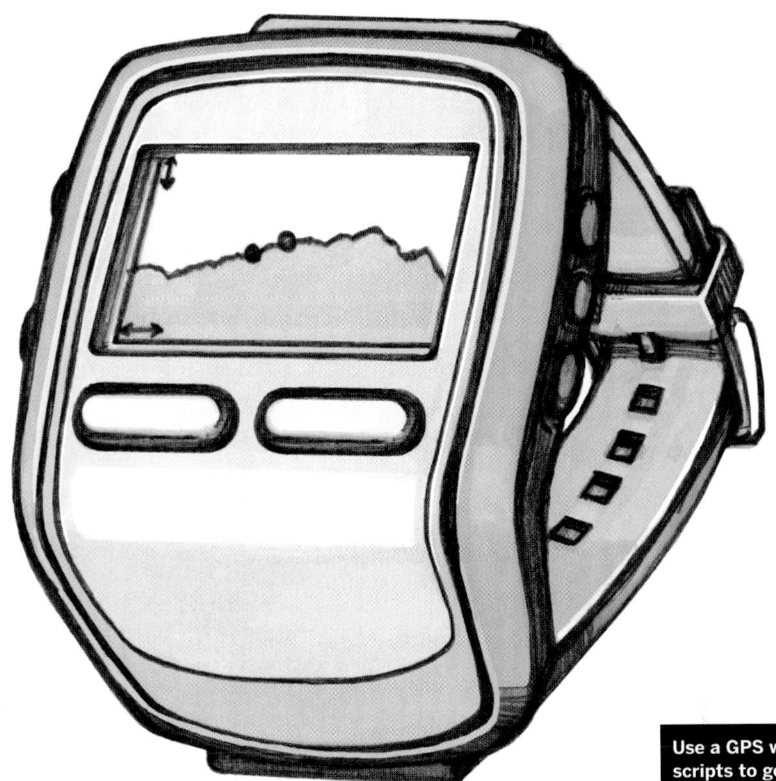

Use a GPS watch and scripts to generate route maps with no effort other than the running itself.

GPS RUNNING LOG

Automatically download exercise routes from a Garmin watch and convert them to animated Google Maps on your website. By Dave Mabe

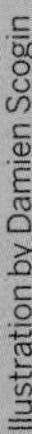

As a technogeek and a serious runner, I love it when my two worlds collide. That's what happened when I started wearing a GPS running watch to log my runs. Previously, I had laboriously maintained a paper log by hand. But the GPS watch let me automate the process with some Perl code.

Around the same time I got the watch, the Google Maps API was released, which allows you to set up a Google Map with custom tracks and labels on your website using JavaScript. So, I also wrote some scripts to display each run as a Google Map on my blog. The race was on!

The watch I use is the Garmin Forerunner 201, which has been on the market for a while, and like many GPS devices, loses reception in wooded areas and near tall buildings. Garmin says its newer Forerunner 205 can get reception in these areas.

Getting and Massaging the Data

The watch recharges and syncs with your computer through a cradle that connects via serial cable. To pull the data off the watch, I used an open source command-line program called GPSBabel (gpsbabel.org), which reads and converts among a variety of GPS file formats. I needed to read the GPS data off my watch in the Garmin logbook file format and then convert it to XML in Google Maps' glogbook format. I used the following command:

```
gpsbabel -t -i garmin -f com7 -o glogbook -F outputfile.xml
```

These flags and arguments tell GPSBabel to read data in Garmin format from the COM7 serial port and write out the data in the glogbook format to an XML file called `outputfile.xml`. The `-t` switch tells GPSBabel that the input will be a tracklog.

Each time the `gpsbabel` command is run, it pulls off every trackpoint that's stored on the watch and saves it to a single XML file. But I wanted to create Google Maps that showed each run individually, not all of them at once. So, I wrote a script in Perl that parses `outputfile.xml`, skips any trackpoint data that the script has already processed, and writes out multiple, separate glogbook files for each run.

To ignore the old data, the script saves the timestamp from the most recent trackpoint it reads and processes only trackpoints that were stored since then.

To separate the new trackpoints into different runs, I defined a threshold, `$MIN_SECONDS_BETWEEN_RUNS`, which tells the script to finish writing the current XML file and start writing a new one when there's a sufficient time gap; in my case, I use a value of 7200, or 2 hours. The script saves XML files for each run and names them using the date of the run (e.g., `2006-01-15.xml`). You can download the code from my website (see Resources).

Google Earth uses an XML-based file format called KML (Keyhole Markup Language), and recent versions of GPSBabel have limited support for this format as well. To create a KML tracklog from an already-saved glogbook file, you run a command following this format:

```
gpsbabel -t -i glogbook -f 2006-01-15.xml -o kml -F 2006-01-15.xml
```

Once I downloaded all the new runs from my watch and ran my Perl script to convert them to both glogbook and KML file formats, I uploaded them to the server where my blog is hosted. I also copied the logs of the most recent runs to files named `latest.xml` and `latest.kml`, so my blog defaults to showing the most recent run.

Generating the Map Pages

To publish Google Maps on my website, I first needed to obtain an API key from Google (each key permits maps to be displayed on a certain URL and all its subdirectories). Then I figured out what I wanted the pages to show. For a rudimentary proof-of-concept, I started with the "Visualize Your Exercise Routes" recipe on Particle Tree's site (makezine.com/go/particletree). Then I modified their JavaScript code to add clickable mile-marker pins that display a pace-per-mile calculation. The mile marker is colored blue if I took longer than 6:30 to run a particular mile, and red if I ran faster than that (you can individualize your own pace-per-mile threshold in the code).

I also added some functionality in the JavaScript to recognize URL parameters that select the run and change the map type from regular to satellite or hybrid. For example, the URL that shows my run from May 10 ends with `index.html?file=2006-05-10.2.xml`. I also used the URL parameters to create an RSS feed of my ten most recent runs.

The JavaScript code uses the AJAX (Asynchronous Java and XML) approach to retrieve the appropriate XML file, parse it, and then map the run trackpoint by trackpoint. The script pauses for a few milliseconds after it maps each trackpoint and then recenters the map, which gives the viewer a dynamic, animated view of the run. No server-side processing is necessary; all the animation action happens in the browser.

Some critical features make the process easy. First, the script knows to end gracefully if the watch isn't cradled. That way, I can schedule the script to run every couple of hours. I simply cradle the watch when it's convenient, knowing that the run data will be automatically uploaded as the watch's battery recharges. A feature I love about the Garmin Forerunner 201 is that you never have to manually clear the run history; it simply deletes the oldest set of trackpoints as it needs more space. Both features allow me to track my running without lifting a finger.

Resources

My running logs: dave.runningland.com/map
Running-log tutorial with downloadable code: dave.runningland.com/grunninglogs

---

Dave Mabe is the author of *BlackBerry Hacks* from O'Reilly Media, Inc., and lives in Chapel Hill, N.C.

A steel draft coil carries beer through its long, cold journey to your cup.

CONVERTIBLE JOCKEY BOX

Portable cooler taps and dispenses ice-cold beer from both kegs and mini-kegs. By Carlo Longino

Photography by Carlo Longino

Going to a tailgater? Here's a homebrew (in the hacking sense) jockey box that dispenses cold homebrew (in the beer sense). A jockey box is one of those funny coolers with a built-in beer tap on the outside. Inside, it has the requisite plumbing to draw beer from a keg, and a metal coil or cooling plate. Once you fill the cooler with ice and attach a keg, the coil or plate, acting as a heat exchanger, quickly chills the beer to a proper serving temperature — even when the keg itself isn't cold.

It's not hard to find pre-made jockey boxes, but they typically cost $150 or more and are "hardwired" with a coupler that works with only one type of keg top. This homebrew solution is not only cheaper, but it's also more adaptable, converting easily into a cooler/dispenser for the 5-liter mini-kegs that many European beers come in. In jockey-box mode, the convertible cooler houses the heat-exchanger coil while the keg stays outside. In mini-keg mode, the entire keg goes inside the cooler, and you don't use the coil. Standard compressed-air coupler and plug fittings let you easily swap internal parts to change the cooler's beer operational mode.

To keep costs down, I used brass fittings from a regular hardware store, but many brewers prefer stainless steel for anything that comes into contact with beer. If you want to go this higher-end route, consult your local homebrew shop.

Assembly

1. If you're using brass parts, remove any surface lead by soaking them in a solution of 2 parts vinegar

MATERIALS

From any homebrew store or austinhomebrew.com:

8+ feet of thick-walled beer tubing 3⁄16" inner × 7⁄16" outer diameter; about 50¢/foot

Barbed ball-lock coupler aka disconnect This fits the 5-gallon ball-lock soda kegs used by home brewers, or you can substitute different hardware to fit standard "D" system commercial kegs. $5

Picnic faucet $4

Mini-keg tap hand pump Mini-keg taps are as cheap as $15; I had a Philtap, which uses CO_2 cartridges and costs about $60 from williamsbrewing.com

Draft coil I used a 50' coil of 3⁄8" stainless steel, available for $75 from morebeer.com (item #H680)

From a hardware store:

1⁄4" × 2" brass pipe nipples (2) About $1 each

1⁄4" × 1⁄4" barbed hose adapter fittings, hose barb to pipe thread (5) About $1 each

Hose clamps To fit the 7⁄16" OD beer tubing (10)

1⁄4" air compressor hose couplers (2) and plugs (3) I bought two Husky brand coupler and plug sets for about $5 each

Drill and drill bits

Adjustable wrench

Rubber mallet

White vinegar

Hydrogen peroxide

Hot water From kettle

Keg of beer With CO_2 dispensing system (or 5-liter mini-keg)

Ice

to 1 part hydrogen peroxide for 15 minutes.

2. Drill 2 holes in the side of the cooler, about 6" apart, below the level of the inside of the lid. Using a 3⁄8" bit, move the drill around to make the holes big enough to snugly accommodate the pipe nipples.

3. Stick the nipples through the 2 holes, using a rubber mallet if necessary. Designate one nipple as the beer-in port for jockey-box mode, and the other as the beer-out. Attach a barbed fitting to the beer-in nipple on the outside and a male compressor plug on the inside. Attach barbed fittings to the beer-out on both sides.

Now it's time for fun with beer tubing. Before each connection, soak the end of the tubing in hot water from your kettle to make it more pliable and easy to get on the barbed fitting. After it's on, secure each connection with a hose clamp.

4. First, make your dispenser by cutting about 18" of tubing, attaching one end to the barbed fitting on the beer-out side, and attaching the other end to the picnic faucet.

5. Next, cut about 4" of the tubing and attach one end to the inside barbed end of the beer-out fitting. Take off one of the compressor couplings, then screw in another barbed fitting and attach it to the other end of the 4" tubing.

6. For the beer-in line, attach a section of tubing (I used about 3') to the beer-in barbed fitting on the outside of the cooler. On the other end of the tubing, attach the coupler for your keg.

7. Cut 2 pieces of beer tubing, about 8" each. Use hose clamps to attach one piece to each end of the draft coil, and insert barbed fittings into the other ends, also securing them with hose clamps. Screw a compressor coupling onto the fitting at one end of the coil (the beer-in) and screw a male compressor plug into the other end (beer-out).

8. The system is now ready to be used in jockey-box mode. Put the coil into the cooler, and hook up the keg and CO_2. Attach the coil's coupling to the plug on the beer-in side, and attach the beer-out coupling to the plug on the other end of the coil.

Fill the cooler with ice, and enjoy your cold draft beer. But remember: You're dealing with pressurized gases and liquids, so exercise due caution.

Mini-Keg Configuration

To convert it into a mini-keg dispenser, remove the picnic faucet from the end of the mini-keg tap. In its place, attach about 6" of beer tubing. Connect a barbed fitting on the other end, screw on a compressor plug, and secure both ends with hose clamps.

Now, all you have to do is uncouple and lift out the coil, tap the mini-keg, put it in the cooler, and plug the coupling into the beer-out port. Cover the keg with ice, and enjoy another cold one.

Resources

Jockey box information: makezine.com/go/jockeybox

Carlo Longino is the publisher of MobileMusicBlog.com and executive editor of TheFeature.com. He can be found in Austin, barbecuing in the dark with a miner's headlamp.

DIY HOME

Out, damned beep! If your child's noisy toys annoy, here's how to shut them up for good.

BEEPKILLER: PARENTAL REVENGE

Three ways to silence annoying toys.

By Erica Sadun

While shopping with the kids, we saw a really cool toy microwave in the clearance section for just a couple of bucks. Being a highly indulgent mother, I sprang for it, not realizing the hidden dangers lurking within: beeps. Lots and lots of annoying, high-pitched, irritating, ear-bending beeps. Push a button, it beeped. The "cycle" ended, it beeped. Ignored for too long, it beeped.

At this point, most people would have simply taken out the batteries and been done with it. But I'm a geek mom, and I don't believe in killing the toy just to kill the sound. So, here are three ways to fix the problem while keeping the toy operational and fun for the kids.

Method 1: Disconnect the Speaker Completely

Difficulty Level: Easy

MATERIALS
Annoying toy
Wire cutters or scissors
Long-tipped Phillips screwdriver

1. Open the back of the toy. Most toys, particularly those manufactured in the Far East, unscrew with a Phillips screwdriver. Make sure to use a screwdriver with a long head because many toys "hide" their screws in long tunnels to keep them away from kids.

2. Locate the speaker. Speakers are round, and you can usually find them near a section of the case that's perforated to let the sound out.

3. Cut one of the speaker wires. It doesn't matter which wire you cut, but leave at least an inch or more of wire next to the speaker (in case you want to try one of the more advanced methods below).

4. Close the case. Return and fasten all the screws to their original places.

Congratulations! You've transformed the toy into a fully functioning but blessedly mute version of itself.

Method 2: Add an On/Off Switch

Difficulty Level: Intermediate

MATERIALS
Annoying toy **Modified per Method 1**
Rocker switch
Wire strippers
Soldering equipment
Hot glue gun

1. Perform Method 1, above. Cut the speaker wire an inch or more away from the speaker, and leave the case open.

2. Strip the wire ends. Use a wire stripper to strip the 2 ends of the cut wire.

3. Identify the rocker switch leads you'll use. Some rocker switches have 2 leads. Others have 6. If you're not sure how your switch connects, read the data sheet. As a general rule, you connect 2 adjacent leads on the same side of the switch (not across the width of the switch).

4. Solder. Tin the ends of your cut wires, and solder them to the 2 adjacent leads.

5. Test the switch. Insert the toy's batteries and test the switch you just installed. The sound should switch between on and off.

6. Glue the switch. Use a hot glue gun to glue the switch somewhere inside the toy.

7. Close the case. Return and fasten all the screws to their original places.

Keep that screwdriver handy so you, not Junior, can open the toy and turn the sound on or off.

Method 3: Add a Volume Control

Difficulty Level: Intermediate

MATERIALS
Annoying toy **Modified per Method 1**
100Ω potentiometer
Wire strippers
Two wires with alligator clips **For testing**
Soldering equipment
Hot glue gun

1. Perform Method 1, above. Cut the speaker wire an inch or more away from the speaker, and leave the case open.

2. Strip the wire ends. As with Method 2, use a wire stripper to bare the 2 ends of the cut wire.

3. Identify the potentiometer leads you'll use. You'll want to connect one wire to the pot's wiper and the other to either one of its end terminals. You can consult the manufacturer's data sheet to find these, but they'll probably be the middle pole and either one of the 2 outer poles.

4. Connect the wires for testing. Connect the 2 wire ends to the 2 potentiometer leads using your alligator connectors.

5. Test the sound's dynamic range. Confirm that you're using a potentiometer with the proper resistance by reinserting the toy's batteries. Hit some buttons to get the toy to make noise. Adjust the potentiometer knob between its 2 extremes and confirm that you can achieve the desired sound level. If the sound turned all the way up is too soft, try swapping in a potentiometer whose maximum resistance is less than 100 ohms. If the sound turned down is still too loud, switch to a pot with a higher maximum resistance.

6. Solder. Once you're happy with the proper potentiometer, tin the ends of your cut wires and solder them to the potentiometer.

7. Glue. Use a hot glue gun to glue the potentiometer firmly to the inside of the toy.

8. Tune the sound. Make final adjustments to the potentiometer's playback levels.

9. Close the case. Return and fasten all the screws to their original places.

Now, you have an internal volume control for your toy. At any time, you can open up the toy and adjust the volume as appropriate for the situation at hand.

Erica Sadun has written, co-written, and contributed to over two dozen books about technology, particularly in the areas of programming, digital video, and digital photography.

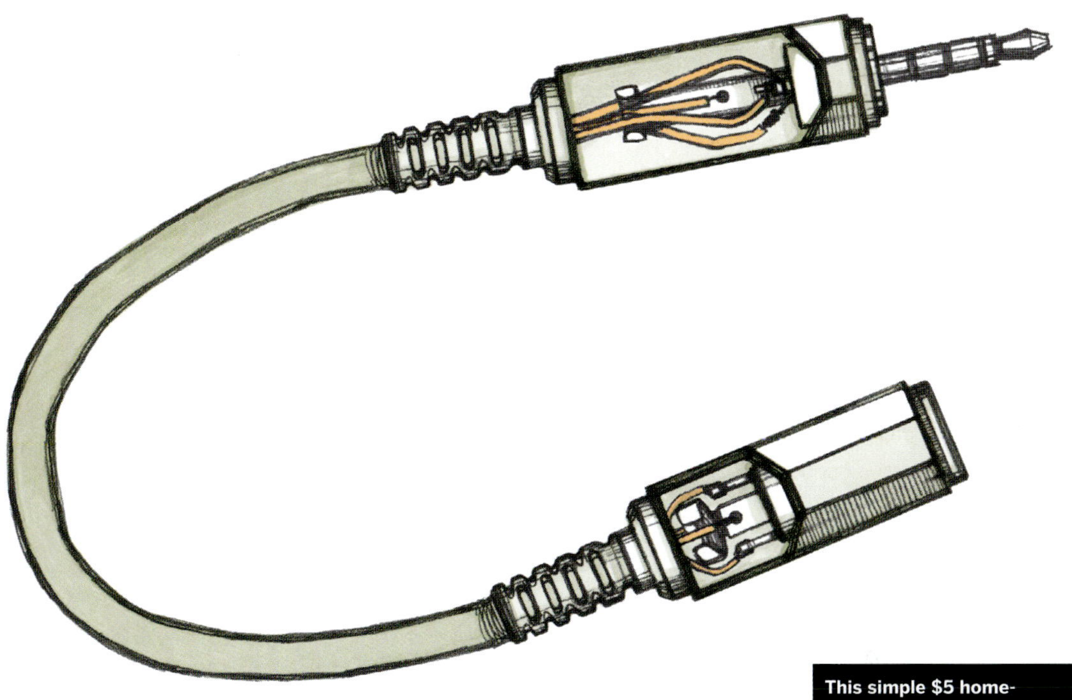

This simple $5 home-made cable lets you hook up your iPod to your television set.

IPOD VIDEO CONVERTER CABLE

An easier way to watch iPod video on your TV.

By Erica Sadun

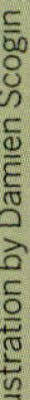

I regularly hook up various portable devices to my TV using a standard ⅛"-to-RCA camcorder cable, the kind with a 3-banded ⅛" plug on one end and a 3-plug yellow-red-white RCA connection at the other. All of these devices except the iPod, follow the standard output configuration.

Apple decided to send the iPod's video through what is normally the right-audio channel (the red plug), the left audio through what's usually designated for video (yellow), and right audio through what's usually slated for left audio (white). Perhaps this is a way of encouraging people to buy the $99 iPod AV Connection Kit.

While it works perfectly well to swap RCA connectors around when you want to feed your video iPod into a television set, it's a pain to have to always reach behind the set and switch plugs around. So I built a converter cable instead. It plugs into the iPod jack and reorders its connections for a camcorder cable that's plugged into a stereo TV the usual way.

All told, the cable cost me less than $5 and took only about 20 minutes to construct and test, once I had the parts. If you've got a spare USB cable lying around the house, it'll cost even less. You'll need to do some wire stripping and soldering, but those are really the only two skills required.

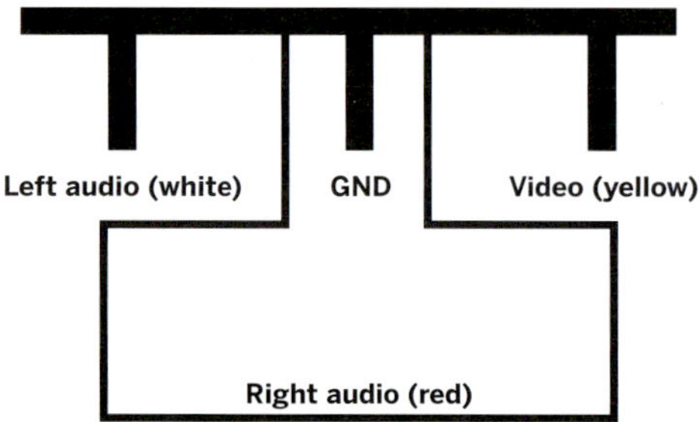

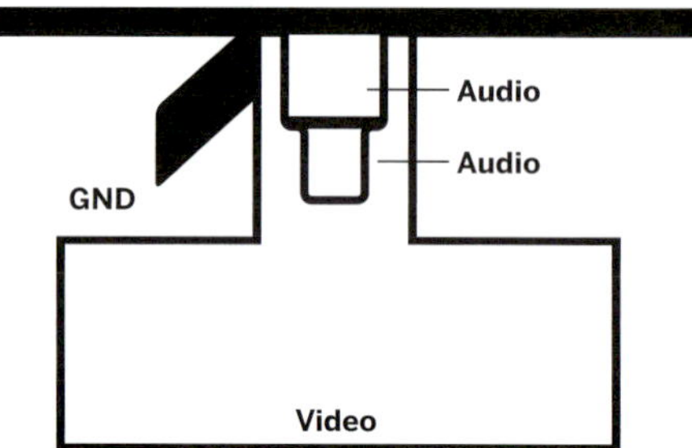

A B

Determining the plug and jack connections. After you unscrew the covers on the plug and the jack, you'll see 4 connectors on each of them. Fig. A: The jack has 3 short rectangles and one large inverted T shaped connector.

Fig. B: The plug also has 4 connectors. Note that the audio channels are on the same plug, but separated by a black line.

The cable uses a ⅛" 4-connector jack and plug — like the iPod, camcorders, cameras, and other devices. This hardware format subdivides the older ⅛" stereo plug into 4 sections that carry right and left audio, plus video and ground.

MATERIALS
4-connector plug mouser.com part #171-7435, $1.53
4-connector jack mouser.com part #161-6435-EX, $1.10
4-wire cable, 4"–8" length You can use a dollar-store USB cable, or any other 4-wire cable will work.
Wire cutter/stripper
Soldering equipment
Multimeter
Standard camcorder A/V cable (not iPod-compatible) For testing

Construct the Cable

Test your connections with a multimeter to be sure you've put everything together properly.

1. Disassemble the plug and jack. Unscrew the black plastic covers off the plug and jack while holding onto the metal of the plug or the gray plastic portion of the jack.

2. Thread the covers onto the cable. Cut any plugs off of the 4-wire cable, then thread on the covers you just unscrewed, back to back, with the thinner part of the covers pointing toward the middle of the cable. The covers are interchangeable, so don't worry about which is which. If your cable is too thick for the covers, trim the constrictive rear sections with scissors or a wire cutter.

3. Strip the wires. Use wire strippers to remove the outer covering from both ends of your cable. Then strip the ends of each individual wire.

4. Identify the jack connectors. The jack uses 4 connectors: 3 short rectangles in a row and a larger, upside-down-T-shaped connector behind them. From left to right, the rectangles are left audio, ground, and video. The big connector in back is right audio. Test the connectors for yourself with a multimeter by plugging a camcorder A/V cable into the jack and measuring resistance between the connectors and each RCA plug.

5. Identify the plug connectors. The plug also uses 4 connectors, but in order to speak iPod, it has a different configuration. (This is not the standard setup.) The short connector hanging off to the left is ground. A small plug in the center contains 2 contact areas for the 2 audio channels, above and below the black line. The large upside-down-T-shaped connector carries video.

6. Solder the jack to the cable. Solder the cable's wires to the jack's connection points. Following the order described in Step 4 (assuming you're using USB wire coloring):

» White connects to the jack's left-audio connector.

» Black connects to ground.

» Green connects to the video connector.

» Red connects to the large right-audio connector.

7. Test your connections. Plug in your normal A/V cable and confirm that the white wire connects to the prong of the white RCA plug, the red to the red, the green to the yellow, and the black to all of the grounds, the circular cuffs around the outside of each RCA plug. If you've misconnected, resolder as needed.

8. Cover the jack. Carefully but firmly slip the black cover nearer to the jack off the cable, and screw it onto the jack.

9. Solder the plug to the cable. Again, assuming USB wire colors:

» Connect the black wire to ground.

» Connect the green wire to the large video connector.

» Connect the white and red wires to the 2 audio connectors, 1 wire per connector.

This last part is the trickiest, as the 2 connectors are so small and close together.

10. Attach the plug cover. As you did with the jack, shift the black cover off the wire and screw it onto the plug. That's it — you're finished and ready to test!

Test It

Connect your A/V cable to your television with the RCA plugs in the normal configuration: yellow plugged into yellow, red into red, and white into white. Connect the ⅛" plug end of the A/V cable to the jack in your new converter cable, and plug the converter cable into your iPod. Play a video out to your television. The audio and sound should match the quality you experience directly from your iPod.

Erica Sadun has written, co-written, and contributed to over two dozen books about technology.

LILIPUTIAN LEVITATION

It has long been known that some materials are repelled by magnets. These materials are called *diamagnetic*. When diamagnetic material is subjected to a magnetic field, the orbits of its electrons change, resulting in a repulsive force.

Water is slightly diamagnetic. In fact, scientists have levitated live frogs by placing them in strong magnetic fields. The element bismuth (which is an ingredient in Pepto-Bismol) is diamagnetic, too. You can explore diamagnetism in a number of ways, but one of the neatest is with a special type of carbon, called *pyrolytic carbon*.

MATERIALS
Four 12mm neodymium magnets
Small piece of pyrolytic graphite
You can buy these from scitoys.com

How to Do It:

1. Slice a thin layer of pyrolytic graphite from your sample. Because pyrolytic graphite is made by depositing thin layers of carbon on top of each other, you can take a single-edge razor blade or a very sharp knife and separate the layers. The thinner the layer, the higher it will levitate. Prepare to make a few mistakes before getting an ideally thin sliver. Your final chip should be a little smaller than the footprint of one of the magnets you're using.

2. Arrange the magnets as shown, with similar poles in the opposite corners.

3. Place the pyrolytic graphite chip flat in the center of the magnet assembly. It should bounce up a millimeter or so above the surface and hover there. If not, you need to slice a thinner chip of graphite.

—Mark Frauenfelder

Further Study:

Demonstrate the diamagnetism of water by using magnets to repel grapes: makezine.com/go/diamagnet

AUTOMATE YOUR VOICEMAIL GREETING

Program Asterisk to daily update your outgoing message in your own voice. By Dave Mabe

I use Asterisk (asterisk.org) to automatically update my own personal voicemail message every day. Asterisk is open source Private Branch Exchange (PBX) software that can turn a Linux box into a corporation-worthy phone system, complete with directory, voicemail, conferencing, text-to-speech, and VoIP features. You can program the Asterisk system with Asterisk Gateway Interface (AGI) scripts, which are written in Perl.

I decided to set up my system to automatically update my voicemail greeting in my own voice, rather than using the cold, robotic-sounding text-to-speech (TTS) engine built into the software. With a little AGI coding and other hacking, I had Asterisk assembling and playing an appropriate, personal-sounding greeting every day.

I used the sound recorder that comes with Windows to record several message chunks: one for each day of the week (*wday1.wav* for Monday, *wday2.wav* for Tuesday, etc.), one for each month of the year (*month1.wav* for January, *month2.wav* for February, etc.), and one for each day of the month (*1.wav* for the first, *2.wav* for the second, *31.wav* for the 31st). I also recorded a greeting start and two different endings: one for when I'm in the office and another for when I'm away. I named these files *start.wav*, *endnormal.wav*, and *endooo.wav*.

By default, Asterisk doesn't have a codec to play *.wav* files. So, rather than install a new codec, you can use the sox sound converter (sox.sourceforge.net) to convert all of your sound files into the *.gsm* format that Asterisk can play. After you convert all your sound files, create a directory called *vm-sounds* in */var/lib/asterisk/sounds* and copy the files into it.

To further automate my system, I set up a motion detector under the desk in my office. I know that if there is no motion in my office between 8 a.m. and 9 a.m., I'll probably be out all day, and my outgoing greeting should reflect this. To monitor the motion detector, I use the excellent home-automation package Misterhouse (misterhouse.net).

I wrote some code for Misterhouse that generates an "office presence" file if I'm in the office between 8 and 9 a.m. The AGI script then checks for the presence of this file when deciding which greeting ending to play. You can get the Misterhouse code at makezine.com/07/diyhome_voicemail.

You can also find the Perl AGI script that puts together the day's message at the above URL. Save it to a file named *vmautomate.pl* and place it in the */var/lib/asterisk/agi-bin* directory. Then add the following lines to your *extensions.conf* file, where 8001 is your phone extension and 100 is your voice mailbox number:

```
exten => 8001,1,Dial(SIP/8001,20,rt)
exten => 8001,2,AGI(vmautomate.pl)
exten => 8001,3,Voicemail,100
```

Finally, reload Asterisk. Now, when you call your extension, your phone will ring for 20 seconds and then the AGI script will run. Depending on the time of day and the presence of the "office presence" file, you should hear the appropriate greeting. You should replace your "regular" voicemail greeting with the default greeting so that it flows properly once it reaches the Voicemail directive.

It's easy to imagine other cool functionalities with AGI scripts. For example, you could take this process a step further and have an AGI script that reads the calendar that you've published with iCal to see whether you're scheduled to be out of the office that day.

Dave Mabe is the author of O'Reilly's *BlackBerry Hacks* and lives in Chapel Hill, N.C.

USB-POWERED FAN

12 easy steps to a cooler you. By Erica Sadun

Google "USB gadgets" and you'll turn up any number of wacky and offbeat USB-powered devices: lava lamps, mini-vacs, instant-noodle cup warmers, smokeless ashtrays. The list goes on and on. It's as if there's an endless demand for USB-powered doodads. Want to know a secret? With USB, you can power nearly any device that normally runs off of two AA batteries.

A standard USB connection provides between 4.5 and 5 volts, and up to 0.5 amps of juice. This means that you can easily convert almost any cool little battery-powered gadget to run on USB power.

Devices that follow the USB standard should (in theory) dock at 0.1 amps, low-power mode, and get permission before switching up to normal power use. But this hasn't stopped all those reading lights, massagers, and other gift items from coming to market, using the USB connection as a simple 5V power supply.

Photograph by Erica Sadun

Here's how you can convert an ordinary battery-powered fan into a USB-powered high-tech masterpiece. You can usually find the main components you need, the fan and a USB cable, at your local dollar store. A little wire stripping, mathematics, resistance, and soldering later, and you'll have your own USB gadget, ready to enjoy.

12-Step Program

1. Divide the USB cable. Cut the USB cable in two. Keep the half with the Type A USB connector, the part that connects to the computer. Put the other half away for another project, another day.

2. Identify the positive and negative terminals. Open the battery compartment and locate the positive and negative battery connections. These

You can convert almost anything powered by two AA batteries into a USB-powered gadget.

MATERIALS

Battery-powered fan You can pick one up for a buck at a dollar store. It should run off of two 1.5V AA batteries.

USB cable Also from a dollar store.

Resistor Ohm value TBD below. If you don't already have the right value resistor, this ups the total materials cost to $3.

Multimeter

Soldering equipment

Wire cutter/stripper

are on the same side as the motor and are usually identified with + and -, with the flat side positive and the side with the spring negative. (You can ignore the other connector pair at the bottom of the compartment. This just serves to connect the two batteries in series.)

3. Measure resistance across the motor coil. Use a multimeter to measure the resistance between the positive and negative terminals. This is the resistance across the motor coil. For example, you might get a reading of 10 ohms.

4. Determine the amperage. Now, some math. Using V=IR (Ohm's Law, voltage = current × resistance), figure out the motor's amperage. You have V and R; two 1.5V AA batteries produce 3V for V, and you just measured the resistance R. So divide V by R to solve for I. For example, a motor with 10 ohms resistance draws 0.3 amps. Note that because USB can provide up to 0.5 amps, this project can only work if the motor's load is over 6 ohms.

5. Calculate the additional resistance needed. Next, determine the resistance you must add into the circuit in order to keep the amps about the same with a 4.5 to 5V power source as it previously had with 3V of battery power. This is another Ohm's Law calculation, but now you use 4.5 for the V, your amperage from Step 4, solve for the resistance, and subtract the motor's existing resistance. With our 10-ohm motor, this gives 4.5V = 0.3 amps × R ohms, or R = 15. With our original resistance of 10 ohms, this works out to 5 additional ohms. Find a resistor with this value.

6. Solder the resistor. Solder the resistor in series (not in parallel) between the motor and the positive battery terminal. To do this, cut the wire that connects the positive terminal to the motor, strip the new ends, and solder them to the resistor.

7. Strip the USB cable. Strip the USB cable to expose the inside wires. There will be 4: power (red), ground (black), data 1 (green), and data 2 (white).

8. Strip and prepare the red and black wires. Fold or cut away the white and green wires. Strip the red and black ones, to expose about ½ inch of the copper threads inside. Twist the threads together for each wire, then tin them with your soldering iron.

9. Solder the red and black wires. Solder the red wire to the positive terminal in the battery compartment, and solder the black wire to the negative terminal.

10. Create an exit hole. Carefully, and in a well-ventilated area, use the soldering iron to melt the plastic on the side of the battery compartment, creating an exit hole for your USB cable.

11. Close up the fan. Run the USB cable through the new hole and close the battery compartment and any other parts of the unit you've opened.

12. Let 'er rip. Plug the fan into your computer, and enjoy the cool relaxing breeze as you admire your workmanship.

Erica Sadun has written, co-written, and contributed to over two dozen books about technology, particularly in the areas of programming, digital video, and digital photography.

INTEREST OF MONEY

Dr. Price, in the second edition of his *Observations on Reversionary Payments*, says: "It is well known to what prodigious sums money improved for some time at compound interest will increase. A penny so improved from our Saviour's birth, as to double itself every fourteen years — or, what is nearly the same, put out at five per cent compound interest at our Saviour's birth — would by this time [1889] have increased to more money than could be contained in 150 millions of globes, each equal to the Earth in magnitude, and all solid gold. A shilling, put out at six per cent compound interest would, in the same time, have increased to a greater sum in gold than the whole solar system could hold, supposing it a sphere equal in diameter to the diameter of Saturn's orbit."

From Burroughs' Encyclopaedia of Astounding Facts and Useful Information, 1889.

Download a copy from manybooks.net/titles/burroughsb1409114091-8.html.

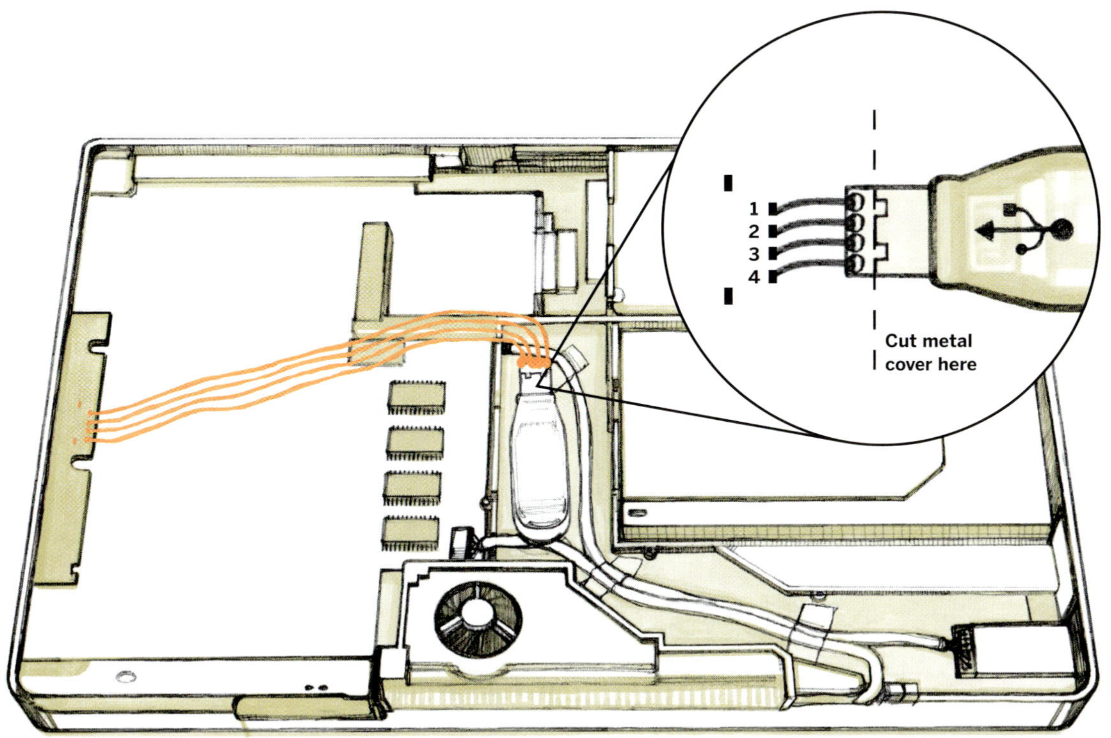

Apple thoughtfully provides the perfect spot to stash your hacked USB Bluetooth antenna.

INSTALLING BLUETOOTH INSIDE AN IBOOK

Tuck a USB wireless adapter inside the case.

By Timothy B. Hewitt

Illustration by Damien Scogin

Apple is kind enough to supply Bluetooth as an option when you purchase an iBook. But if you want it later, you are S.O.L. — they won't sell you all the parts you need to make the conversion. Luckily, it's not difficult to do your own internal Bluetooth installation by wiring one of the USB port's contact points to a USB Bluetooth adapter hidden inside the laptop's case.

After getting a new Bluetooth-enabled phone, I wanted to use it with my iBook. So I bought a few Orange Micro Blue2 USB units from eBay (they were cheap enough to buy some for me, some for gifts). But after the novelty wore off, I started thinking that the antenna unit looks stupid sticking off the side of the iBook, and you always have to carry it along and plug it in, just like you would with a cable. This begs the question: What's the point of having it be wireless?

I'd had enough. I really wanted internal Bluetooth, and I wondered if I could fit the adapter inside the iBook somewhere. The answer turned out to be yes.

Some Disassembly Required

I started by taking the iBook apart to see what space, if any, was available inside. If you haven't taken an iBook apart before, visit PowerBook Tech (powerbooktech.com). They have detailed instructions on ripping Apple Books apart. Note

MATERIALS

- **Apple iBook** That lacks factory-installed Bluetooth
- **Bluetooth USB adapter** Such as the Blue2 from Orange Micro
- **Wire-wrapping wire**
- **Electrical tape**
- **Dremel tool**
- **Small screwdrivers**
- **Multimeter**
- **Soldering equipment**

that if you are going to try this, you do so at your own risk! You could seriously damage your lovely Apple product.

I originally figured that I would have to crack open the Bluetooth adapter and install just the tiny chip inside of it somehow. But a look inside the iBook showed that there was enough space for the entire adapter. I found the perfect spot behind the CD/DVD drive, near the fan.

Also note that there are a few new screws on these models, as compared to the ones shown on PowerBook Tech. Especially note the 3 next to the CD/DVD drive. Keep track of these. I always somehow end up with one left over.

Hack the Port

If you are looking at the bottom of the logic board with the USB ports on your left, you'll see that each USB port has 6 solder points. The 2 outside points are power and ground, and are used to secure the port. The 4 tiny inside points are the actual connections you need to make. The illustration (on previous page) shows how they line up, facing the bottom of the logic board with the USB ports on your left.

It turns out that you can get to the solder points for the iBook's USB ports without all of the unscrewing around, but I wanted to get a meter in there and be sure that I had the correct connections.

After finding where the points were, I took my handy Dremel to the plug on the Bluetooth adapter to remove some of the metal plug and expose the connections, as shown. I only removed a little bit of the metal; I didn't have to go too far and didn't want to risk cutting the connections.

For the connections to the USB contacts, I went to an electronics supply store (Main Electronics here in Vancouver) and asked for the thinnest wire they had. They recommended Wire Wrapping Wire, which they said "the kids use for mods to PS2s and things." Good enough for me!

This wire is thin enough to run over the logic board without taking up space, and the case cover fits over it just fine.

I soldered 4 wires to the Bluetooth unit, then connected them to the 4 corresponding contact points on the logic board. I didn't need to use any additional solder on the contact points, since the wires were so fine.

I wrapped the connector end of the Bluetooth unit in electrical tape for insulation, and set it in the space I'd found for it. I then taped it in place, just like Apple does with its cables. A quick boot showed I had Bluetooth, so I screwed it all back together.

The only thing "bad" about this mod is that you can't use the hacked USB port for other devices. I was hoping things would be daisy-chainable, but no luck.

Timothy B. Hewitt is a longtime Mac user who can't leave well enough alone.

WONDERS OF MINUTE WORKMANSHIP

In the 20th year of Queen Elizabeth, a blacksmith named Mark Scaliot made a lock consisting of 11 pieces of iron, steel, and brass, all which, together with a key to it, weighed but one grain of gold. He also made a chain of gold, consisting of 43 links, and, having fastened this to the before-mentioned lock and key, he put the chain about the neck of a flea, which drew them all with ease. All these together, lock and key, chain and flea, weighed only one grain and a half.

Oswaldus Norhingerus, who was more famous even than Scaliot for his minute contrivances, is said to have made 1,600 dishes of turned ivory, all perfect and complete in every part, yet so small, thin and slender, that all of them were included at once in a cup turned out of a peppercorn of the common size. Johannes Shad, of Mitelbrach, carried this wonderful work with him to Rome, and showed it to Pope Paul V, who saw and counted them all by the help of a pair of spectacles. They were so little as to be almost invisible to the eye.

From Burroughs' Encyclopaedia of Astounding Facts and Useful Information, 1889.

Download a copy from manybooks.net/titles/burroughsb1409114091-8.html.

THUMB LIFE

USB keydrive lets you listen to, read, and play what you want on any machine. By Russ Ethington

At first, the USB keydrive looked like a replacement for the outdated floppy disk. But new uses for the keydrive are being invented all the time, and an inexpensive keydrive lets you take along music, volumes of literature, a reference bookshelf, and games everywhere you go. This article describes some fun multimedia adventures you can take with your keydrive. But first, let's look at how it works.

What Makes This Little Gizmo Go?

Inside its simple exterior, a keydrive is a sophisticated little package of electronics and software. Its three main components are a flash memory chip, a microprocessor, and a USB adapter. The flash memory stores the data. The microprocessor continuously runs a program that reads and writes data to the chip and imitates a hard disk drive to any device the keydrive connects to. The USB port makes the physical connection and supplies enough power to run the memory chip, microprocessor, and data communication lines to your computer. Thanks to the built-in drivers on modern operating systems, the keydrive's interaction with the OS is transparent; it mounts to the file system immediately, and looks just like a regular disk drive to almost any software you run.

Hack #1: Portable Music Player to Use with Any Computer

You can turn a keydrive into a poor man's take-along music player. It'll be smaller than an iPod Nano, but you need to plug it into a computer to play it. First create a */Music* folder on your keydrive and fill it with MP3 audio files. On a 512MB drive, you can expect to fit about 100 tunes, or about 8-10 hours of music.

Then you'll need to add software to organize playlists and play the songs when you plug in the keydrive. Good software for this depends on your platform. Make sure you save both the player software and your configuration file onto the keydrive. For example, with Zoom Player (I used the free Standard version), create a directory on your keydrive called */ZoomPlayer*, and install the app into this directory. Run the program from its icon on the keydrive, configure it to your liking, and organize your playlists. Then bring up the configuration dialog (Options → Advanced Options → Settings → Other) and select "Save configuration in a local file." This ensures that your keydrive will remember your settings, playlists, and even where you left off listening last time.

Windows

Zoom Player: A small, feature-rich player
inmatrix.com/files/zoomplayer_download.shtml
XMPlay: Another sleek and capable player that supports many sound formats un4seen.com
Winamp: Probably the best-known player for Windows, now a portable app
winamp.com/player/free.php

Mac OS X

iTunes, Quicktime, and Preview: Already included in OS X (so there's no need to copy software to the keydrive)
MP3 Rage: MP3 organizer and player
chaoticsoftware.com/ProductPages/MP3Rage.html

Linux

XMMS: Already included in most Linux distributions, but just in case, take along a binary:
xmms.org/download.php

Hack #2: Giant Literature Bookshelf

A keydrive can carry a terrific collection of important literary works, for whenever you need something great to read.

Visit Project Gutenberg at gutenberg.org/catalog and enter a simple search, for example "Shakespeare," in the Author field. Results will include all of the bard's works listed individually, and an entry for "The Complete Works" (EText #100). Click on one of the plain text links for The

Complete Works, save it into a */Books* directory on your keydrive, and you'll have the whole enchilada. Next, why not grab some Twain, Dickens, or Tolstoy? It's easy to create a free, giant bookshelf of classics.

If you have room, try Gutenberg's "Best Of" canon of world literature, EText #11220. This includes over 600 works — a lifetime of reading — but you'll need at least 750MB of space.

Every Gutenberg book is available as plain text, so you can read them in your favorite browser or text editor without special software. Many are also available formatted in HTML, Plucker, DVI, EPS, and other formats, for easier reading.

Hack #3: Reference Library

You can apply Hack #2 to put a wealth of reference information at your fingertips, in addition to literature. Here are 12 reference titles available from Project Gutenberg, listed along with their searchable EText numbers:

» *Webster's Unabridged Dictionary*, #673
» *Roget's Thesaurus*, #22
» *Project Gutenberg Encyclopedia* (first volume), #200
» *The Nuttall Encyclopedia*, #12342
» *CIA World Factbook*, #6344
» *US Census Figures Back to 1630*, #115
» *Bartlett's Familiar Quotations*, #16732
» *Toaster's Handbook: Jokes, Stories, and Quotations*, #12444
» *Handy Dictionary of Poetical Quotations*, #15119
» *United States Constitution* (without Amendments), #5
» *United States Bill of Rights*, #2
» *Dewey Decimal Classification*, #12513
» *Etiquette*, by Emily Post, #14314

Hack #4: A Retro Gaming Device

Along with a bit of determination, the free game machine emulator called MAME opens up a world of portable gaming, letting you run classic arcade-style games on any computer. First, download and install the MAME software on your keydrive. Again, which software depends on your platform.

Windows
MAME mame.net/downmain.html

Mac OS X
Mac MAME macmame.org/downloads.html

Linux
Advance MAME advancemame.sourceforge.net/download.html

MAME enables your computer to emulate any of a large collection of gaming devices. Download and install MAME onto your keydrive. Then you need to add the games themselves, which are available online as ROM files. Some are in the public domain, others are for sale, and most are protected by copyright — so be sure of your source. Here are a few places to begin your search:

» MAME main site ROMs mame.net/downmisc.html
» StarROMS authorized reseller starroms.com
» ROM World rom-world.com

As with the portable music player, you now have your portable application and your content files. Just plug and play!

Windows users can also download games that run standalone, without MAME, at the Portable Freeware Collection, listed below.

More to Explore: Portable Applications

In Windows, typical computer programs must be installed with a special setup program before you can use them. A better approach, used on Unix and the Mac for decades, lets you install applications by simply copying or unpacking a single file or directory. Thanks to the keydrive's popularity, many Windows developers are using this approach to create new portable applications. Great open source programs like Firefox, OpenOffice, and Thunderbird are now available as portable applications that can be easily keydriven from one computer to another. You can find good portable applications at these sites:

» Portable Freeware Collection: portablefreeware.com/about.php
» U3 portable application platform: u3.com
» Portable (open source) apps: usbapps.com
» Free Open Source Software Mac User Group freesmug.org/portableapps

---

Russ Ethington is a software developer by day, inventor by night, and occasional freelance writer who likes to take things apart to see what makes them tick.

EL CHEAPO CANTENNA

"Mountain Grown" coffee can makes homegrown wi-fi range extender.

By Will O'Brien

The cantenna is a venerable hack that uses an empty metal can to extend the range of your wi-fi network. The idea is that if you connect a regular wi-fi antenna to a can, it will focus the signals in the direction of the can's open end. Cantennas work at either end of the connection; you can keep a computer networked from farther away by using a cantenna pointed toward the direction of an available wireless router, or enable a router to feed otherwise out-of-range computers by pointing a cantenna toward them.

There are many cantenna designs out there, but this one is the cheapest. All you need is a coffee can and a large soldering iron, and you can extend the range of any wireless card that has an external dipole antenna. I used Belkin's Wireless Desktop PCI Network Card 802.11b, part #F5D6001, which has a dipole at the end of a short pigtail and costs only $5 from CompUSA. Unlike other cantenna designs, this one does not require any expensive N connectors or extra pigtails.

The inner wire of your card's antenna, the signal half of the dipole, pokes through the side of the can. For best results, you need to position it carefully, with its optimal depth and location derived from the can's diameter and the wavelength of the wi-fi channel you're using. Greg Rehm's wi-fi page (turnpoint.net/wireless) has a handy calculator that figures these numbers for you. For my

Directional cantenna lets distant wireless network clients wake up and smell the wi-fi.

Photography by William O'Brien

The Belkin card's external antenna has a plastic cover that comes off easily, revealing the dipole inside. To assemble the cantenna, just solder the outer part of the dipole to a hole drilled in the side of the can and let the inner wire stick straight into the center. Then add the mount of your choice.

4"-diameter coffee can, the dipole element needed to be 1.12" tall, and placed 2" from the back of the can. When I hacked into the cheap Belkin card and measured the inner wire of its antenna, I was happy to find it was just about the right length.

Build It

1. Acquire a 13-ounce coffee can, or similar metal can (4" diameter).
2. Drill a 3/16" (or so) hole, 2" from the back of the can, measuring from the inside, not the outer lip. The hole should be about the size of the dipole antenna you're modifying.
3. Cut the plastic covering off the antenna, to reveal a metal tube surrounding an inner wire. These are the 2 elements of the dipole.
4. Insert the antenna into the hole so that the inner wire sticks in about 1 1/8". Use rosin-core solder to connect the antenna's metal tube, the shielding element, to the can. Solder from outside, making a good, thick solder bead around the edge of the hole. I used "helping hands" alligator clips to keep the antenna in position.
5. Add a mount of your choice. I made a simple one by cutting up the original antenna cover and adding a zip tie.

NOTE: You can use this same design on the router side. My Linksys WRT54GS has 2 dipole antennas, and you could cantenna-fy it by unscrewing one of them and substituting a can-enhanced pigtail antenna with a Linksys-compatible connector.

Will O'Brien pulls espresso and modifies innocent kitchen appliances somewhere in middle Missouri.

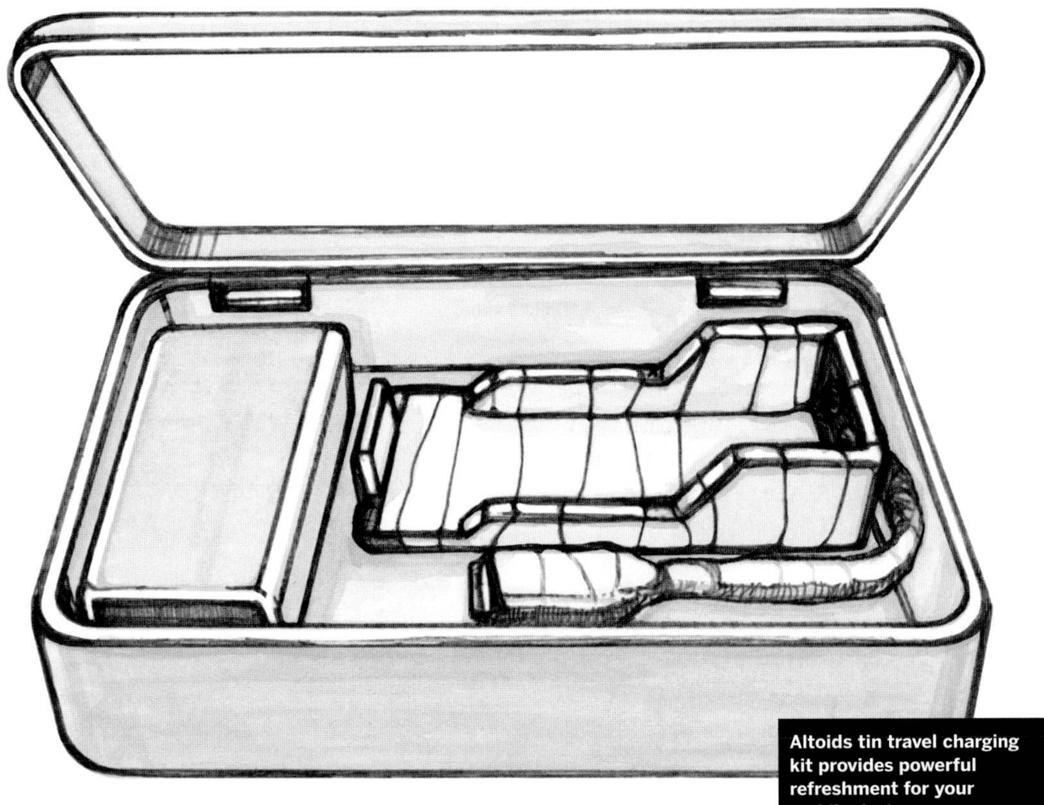

Altoids tin travel charging kit provides powerful refreshment for your mobile devices.

BACKUP POWER TO GO

9V battery USB-compatible charger juices up portables in a pinch. By Erica Sadun

Illustrations by Damien Scogin

You're at a café, reading your email or updating your calendar on your PDA, and *bling*, the battery warning starts flashing. "You're running extremely low on battery power." You stop what you're doing, power down your device, and eloquently curse that engineer at Dell who thought that two hours was a reasonable amount of time for a PDA to be in operation. Sure, you could have carried the cumbersome power cord and adapter with you, but that defeats the whole purpose of not using a laptop. Remember the bit about how you can carry the PDA around in your pocket?

And it's not just PDAs. Cellphones, iPods, and other useful little devices insist on being docked to a wall so they can recharge their way back to usefulness. Wouldn't it be easier if you could run down to the drugstore and pick up a few standard-size batteries to run your gadgets off of? Fortunately, you can. Here's how you can put together a simple USB-compatible battery-driven power source to extend the working life of your portable devices. It's easy: no soldering, no breadboards, and not a lot of fuss.

The secret ingredient is a voltage regulator, which is an integrated circuit that converts a wide range of input voltages into a constant, regulated output voltage. With a voltage regulator, you can stick a 9V battery at one end and produce gadget-friendly, USB-alike 5V output at the other.

This article shows how to use a voltage regulator to produce a 5V power source for any USB-powered device, but you can adapt the same steps to create a portable power source for nearly any device. You just need to set the right output

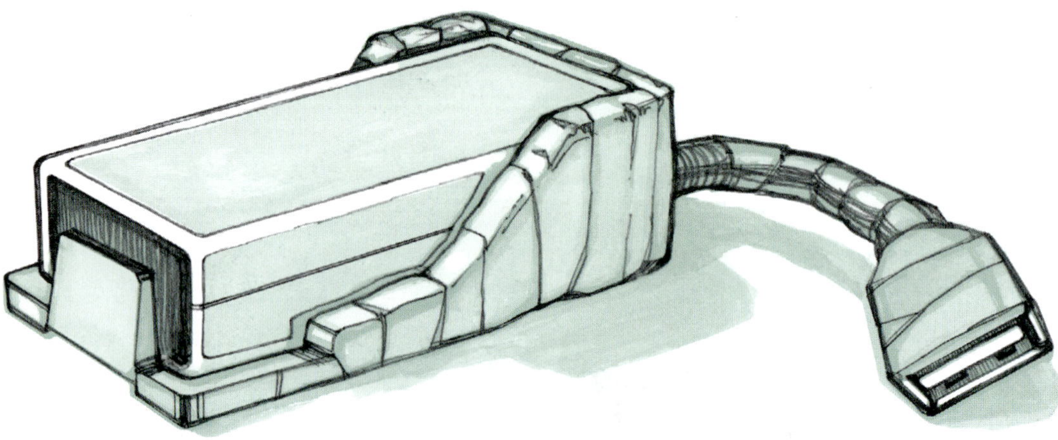

Homemade voltage-converting charger is ready to give a USB power transfusion from a donor 9V battery to a dying PDA recipient. The key component is a 7805 voltage regulator at the base of the cable, which takes the battery's 9V down to USB-standard 5V.

MATERIALS

- **LM7805 voltage regulator** From your favorite geek/ electronics shop, or 42 cents from mouser.com (part #512-LM7805CT), or $1.59 from RadioShack (part #276-1770)
- **Female USB connector** Also from geek shop, or 56 cents from mouser.com (part #154-2742), or cut one from a spare USB cable
- **Fresh 9V battery**
- **9V battery connector** Either a simple cap (about 10 cents) or a holder style (about $1), with attached leads
- **Insulated wire** 12-gauge or so
- **Electrical tape**
- **Project box (optional)** I use an Altoids tin to store the power supply and the battery. It smells nice and adds that certain je ne sais quoi.
- **Multimeter**
- **Wire cutters/strippers**

voltage and configure it to take enough batteries to produce at least 2V above the target output voltage. For example, you can make a 12V charger that uses two 9V batteries.

NOTE: Many 5V power supply project sheets available on the internet call for capacitors and diodes to prevent damage to the voltage regulator from spiky or reversed input voltages. After consulting with physicist Randall Tagg, I determined that this project doesn't need either of these components. A 9V battery's output is steady enough not to require smoothing by a capacitor, and anyone who is sufficiently motivated to forcibly plug a 9V battery the wrong way into a battery cap probably deserves whatever results he or she gets.

Charging Ahead

In the following steps, all your connections are made with electrical tape rather than solder. This simplifies construction and produces a light, flexible result.

1. Trim 2 wires to about 4" each. Strip their ends. You'll use these wires to connect the USB connector to the output of the voltage regulator.
2. Connect the positive lead from the 9V battery holder to the positive input terminal of the LM7805 voltage regulator. The input terminal is at the left when the chip is face-up.
3. Connect both the negative lead from the 9V battery holder and one project wire to the common (ground) terminal of the LM7805 regulator. This

is the middle terminal. Make sure both wires make good contact with the terminal when you tape them up.

4. Connect the second project wire to the output terminal of the LM7805 regulator. The output terminal is on the right.

5. Test the output. Insert a 9V battery into the holder and use the multimeter to test the output. It should read a steady 5V. Remove the battery after you've confirmed that the output works as expected.

6. Connect the ground wire from the regulator to the USB connector ground. This is pin 4 on the connector, the leftmost terminal when looking at the connector from the back, which connects to the leftmost terminal when looking into the plug, with the USB symbol face-up. The pins on the connector are small and delicate, so take care when making your connection with the electrical tape.

7. Connect the positive wire from the regulator to the USB connector. The positive wire connects to the rightmost terminal, pin 1, again looking at the connector from the back.

8. Wrap everything in electrical tape. Use electrical tape to create a finished presentation, wrapping the wires and the back of the USB connector. Do not, however, cover the IC portion of the voltage regulator. The regulator sheds its extra voltage as heat, so you should leave the flat box exposed to open air.

9. Try it out. Attach the 9V battery, plug in your favorite gadget, and watch in wonder as it recharges itself from the battery. Afterwards, detach the gadget and remove the battery.

You're now ready to put your new, ready-to-travel USB power source and a 9V battery into your Altoids tin. If you want to get fancy, you can screw the voltage regulator to the tin itself to take advantage of the flat metal surface, which will work as a heat sink. Spare power is ready and waiting. And remember to always remove the battery when not in use, or you'll drain it.

Erica Sadun has written, co-written, and contributed to over two dozen books about technology, particularly in the areas of programming, digital video, and digital photography.

Emergency USB Cable Charger

On my last annual trek to the lovely Outer Banks in North Carolina, I managed to leave my cellphone charger at home. It was a 20-hour car ride, and turning back was not an option. Worse, my Danger Sidekick has an internal battery that's good for only a couple of days, and I was going for a full week.

Lucky for me, our rest stop just south of Norfolk, Virg., was next to a RadioShack. I didn't feel like paying for a charger that I would use for only a few days, so I decided to make my own simple USB cellphone charger cable for just a few dollars' worth of mostly reusable parts. I shopped for the ingredients I needed, warded off the overly helpful employee with a counter-curse, paid for the bits of my potion, and was ready to go. I put it together in a few minutes.

This charger works for any phone that takes 5V to charge, which is common. Your phone's wall charger will be labeled with the voltage it supplies.

MATERIALS

- USB cable with USB-A plug **Either A-to-A or A-to-B**
- Replacement Adaptaplug socket **RadioShack**
- Size H Adaptaplug **RadioShack**
- Electrical tape
- Nylon wire ties
- Multimeter

1. Cut your USB cable, and separate and strip the wires on the male "A" plug end.

2. Plug the cable into your computer, being careful not to let the wires short. Use a multimeter to test pairs of wires and identify which are the +5V and which are the ground. (On my cable, red was +5V and black was the ground, according to the standard.) After you've identified the wires, unplug the cable.

3. Connect the replacement Adaptaplug socket to the USB cable, joining the TIP (the wire to the plug's tip, with the printing on it) to the USB +5V and the base to the USB ground. Use a wire tie to prevent the wires from pulling apart and insulate any bare wire with electrical tape.

4. Plug the "rather expensive" ($4.99) "H" Adaptaplug into the socket, with TIP connected to the + side.

5. Plug the USB cable into your computer and check the voltage at the new plug with a multimeter. If you don't get a clean, steady 5V, you may not want to risk your expensive cellphone by plugging this in.

Fortunately, mine was dead-on at 5V, so I went ahead and plugged in my Sidekick. I was rewarded with the battery-charging indicator. Before I crashed into bed, I made sure Energy Saver mode on my laptop was turned off, in order to keep the USB bus powered overnight. The charger worked perfectly for my entire trip and now lives tucked away in my computer bag just in case.

—Will O'Brien

Book-bound PDA computer has elegance, versatility, and minimal power consumption.

PALM PILOT NOTEBOOK

Modified hardback book contains extra-powered PDA and travel keyboard. By Allen Wong

My ideal laptop is an instant-on ultraportable with flash memory and battery life measured in days, not hours. After a PDA upgrade left me with a perfectly serviceable Palm IIIe, I decided to build what the market couldn't deliver. A trim 1.1lb., this instant-on laptop uses standard batteries and runs basic apps on the Palm OS. Made of 90% recycled content, it is low-energy, low-emission, low-cost, and user-friendly. Garbed in the technology of the past, it portends devices of the future.

Construction

1. Disassemble and discard Palm Pilot case. The essentials of my Palm IIIe were a screen stacked on a processor board, connected by a short ribbon cable. Also detach the keyboard's cover and cradle (but keep the cable leading to the cradle connector). You don't need any excess injection-molded fat.

2. Arrange the parts. Lay out all the functional bits in a way that approximates the low-powered ultra-portable of your dreams. I separated the PDA screen from the board to make the cover side thinner and put the serial cable and extra batteries above the keyboard. Your arrangement will vary depending on the PDA and keyboard you start with. Visualize the book covers and where the split will be. Think through where you'll place all the buttons and ports (including IR and card slots), and how you'll route the cables.

3. Select a book. You can scour your collection, or look for books with less emotional attachment at garage sales, thrift shops, and library sales. Bonus points for a cool title or great

Photography by Allen Wong

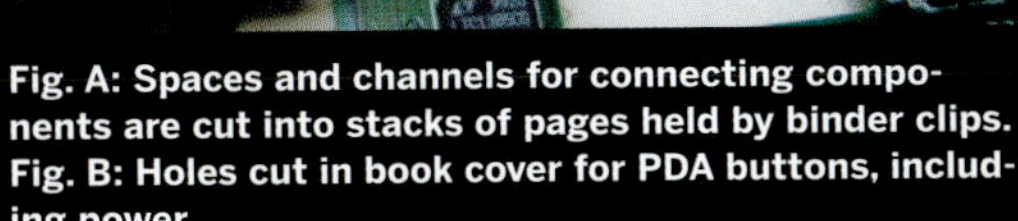

Fig. A: Spaces and channels for connecting components are cut into stacks of pages held by binder clips.
Fig. B: Holes cut in book cover for PDA buttons, including power.

Fig. C: If you're using a sync cable, refer to its connector for how to solder the leads to the PDA's board.
Fig. D: All connections to the PDA's main board; 3 for the keyboard at left, and 5 for the cable, in the middle.

MATERIALS

- **PDA** I used an old Palm IIIe (2MB). Any PDA will do, although some are easier to open, have fewer parts, and won't be missed too much if accidentally drilled through.
- **GoType! keyboard** from LandWare (landware.com), or another third-party PDA keyboard
- **Some way of syncing the PDA** I used a serial hotsync cable, but its size caused problems. If I were to do it again, I'd find a thin USB cable or use Bluetooth, wi-fi, or infrared to sync.
- **Triple AA battery holder**
- **Hardback book**
- **Length of ribbon**
- **Steel ruler**
- **Hobby knife and lots of blades**
- **Drill**
- **Needle files**
- **Binder clips**
- **White glue**
- **Double-stick tape**

cover art behind the jacket. Most importantly, the book must be wider, taller, and thicker than the arrangement of your parts. How close you cut it depends on your skill with a hobby knife and how much you value portability. In my quest for extreme portability, I chose a book that left parts visible from the sides. My serial connector was so thick that the book wouldn't close until I filed down its sides. If you want the parts to be completely hidden when the book is closed, use a bigger book.

4. Remove and divide the pages. Using a steel ruler and a hobby knife, cut out all the pages about ¼"-⅜" from the spine. You'll need these pages later, so the less tearing and shredding, the better. Divide your stack of pages into two: one stack for the keyboard, the other for the screen. The height of the stack will depend on the height of your parts. Reserve a few blank pages, such as the end leaves, for later use.

5. Trace the parts and cut the pages. Position the parts on these stacks of pages, tracing them in pencil, and redraw the lines to make them as straight as possible. (Fancy curved cuts will only cause you grief.) When cutting, use a steel ruler with a cork or rubber backing and sharp, new blades. Cut only a few pages at a time, rather than hacking through the entire stack all at once. Visualize the path of the blade, and make sure nothing blocks that path. Please, be safe!

Make hidden channels for cable runs by cutting only the pages well underneath the top. You might find that certain components on the printed circuit board are deeper than others, and I cut many extra "wells" so that the boards would lie flat. It's easier to subtract than to add paper, so try to get it right the first time so that the parts fit snugly and won't rattle around. For additional depth, you can add extra pages beneath your current piles. Work in layers and use binder clips to keep loose stacks organized.

6. Cut the cover. After I splayed the Palm Pilot screen from the board, all of its buttons (including the power button) faced the book cover. To provide access to these buttons, I used the case as a stencil to trace the buttons on the cover, and then drilled out the centers and used needle files to shape the holes.

7. Connect the electronics. Test-fit all of the electronic parts in your paper piles. Note the length of any connecting wires and make sure paths are cut for all connections. If wire-stripping and soldering aren't your strong suits, make the wires a little longer so you'll have a few extra tries.

First, connect the keyboard to the PDA by cutting the wires leading to the keyboard's cradle connector and soldering them to the corresponding traces on the Palm's main board. My keyboard used only 3 wires, which extended from a small auxiliary board. Save the cradle connector so you know where to solder the wires.

Then connect the sync cable, if you're using one. As with the keyboard, cut off its connector and refer to it as you solder the cable's wires directly to the contacts on the Palm Pilot main board. Mine used 5 contacts.

My sync cable joined the keyboard wires in a huge soldery mess. The whole serial cable business was a lot of extra work.

Finally, hook up power. My PDA's battery compartment was molded into its case, so I cut out the battery springs and soldered the leads to an external battery holder. After everything's soldered, fire up the parts to ensure that they all work and the connections are strong.

8. Glue the pages together. Start from the edges using slightly watered-down Elmer's glue. Brush it on the edges of the stacked pages and immediately clamp them under stacks of books so they don't curl. Give it plenty of time to dry. Plastic

The PDA book computer is powerful enough for writing and other word-processing tasks, and distinguishes you from all the laptop clones at the coffeehouse.

wrap placed above and below your parts will help keep them from sticking.

9. Attach the ribbon. A length of ribbon keeps the screen at the proper viewing angle and prevents the laptop from tipping over. It's tricky to find that exact balance point before everything's assembled. First, attach the ribbon to the "screen" side of the book, and leave the excess hanging over the top. Then, attach the screen assembly to the lid. Find that perfect screen angle, and tape down the free end of the ribbon. Attach the keyboard assembly securely over the ribbon so it won't slip out. Use strong double-stick tape to stick the pages to the covers.

10. Finishing touches. Remember that blank end-leaf paper you saved? It's time to make your last precision cut and create a paper screen bezel. This will be the most visible part of your laptop, so make it nice! Trace the Palm Pilot's screen bezel and then straighten the lines with the ruler. Double-stick the bezel over the screen.

Congratulations! You've built your own piece of recycled retro-geek-chic! Decorate the outside with stickers or the book jacket of your choice, and enjoy the extra table space at the coffeehouse.

Allen Wong is an idea guy who lives in Los Angeles and spends his spare time inventing for a parallel universe.

Stepper right up: Webcam-software-plotter system draws your likeness While U Wait.

SCRIBBLER BOT

Homemade two-axis plotter finds work as a caricature artist. By Douglas McDonald

Photography by Douglas McDonald

The Scribbler Bot is a software-hardware system that converts a digital image into a stylized physical drawing. The software component converts a webcam snapshot into a line drawing that consists of a dense network of "scribble lines." It then exports this vector-based representation as a set of point-by-point motor commands to the system's hardware component: a homemade orthogonal plotter. The plotter creates the drawing using a pen, pencil, or any other medium you can attach with a zip tie.

This article focuses on building the plotter, which moves a stylus along both x- and y-axes — in contrast to a printer, which moves its print head along one axis. Orthogonal plotters, aka flatbed plotters, are often used for full-sized engineering and architectural drawings.

Plotting the Course

The first step in building the Scribbler Bot was finding a way to control and move the stylus along two axes. I needed a system that could keep track of where on the page the stylus has been, is currently, and needs to go, while correlating these locations to their analogous positions in the image's digital representation.

Creating this setup is possible by using stepper motors — which move in discrete, programmable amounts — and a dual-motor control system. After some searching, I decided to use the A200-SMC system from Stepper Control (steppercontrol.com). The A200 includes a hardware controller board that connects to a Windows machine via parallel port, and software (which I did not use) that commands the controller board from a PC.

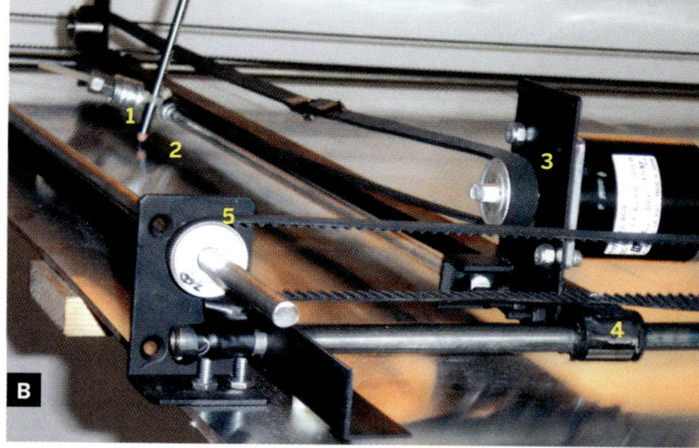

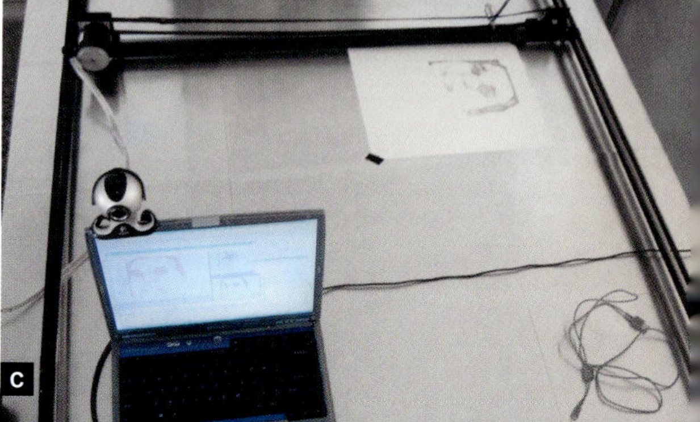

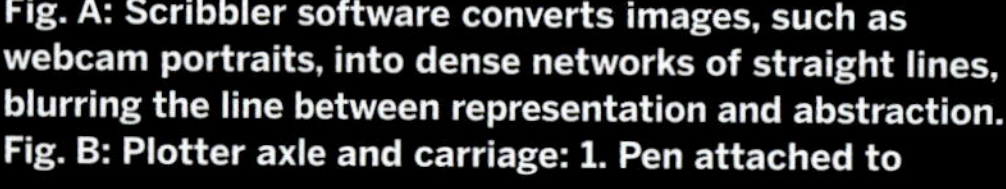

Fig. A: Scribbler software converts images, such as webcam portraits, into dense networks of straight lines, blurring the line between representation and abstraction. Fig. B: Plotter axle and carriage: 1. Pen attached to stylus bearing with zip tie; 2. Stylus bearing on carriage rod; 3. Carriage motor and belt; 4. Main belt, bearing, and track; 5. Main belt pulley and axle. Fig. C: Full system, with webcam and software on laptop.

For the motors themselves, I just used generic Chinese models.

The Scribbler Bot software (see sidebar next page) is written in Macromedia Director. Conveniently, Director also has an ActiveX control that handles parallel port I/O. Embedding this control into the Scribbler project software lets it control the A200 board directly, through Director, rather than having to go through the A200's own separate controller application.

Building the Structure

The roughly 42"×40" drawing area of the Scribbler Bot is framed by the main tracks, which define the plotter's x-axis. The tracks consist of two smooth metal rods that run along the top and bottom edges. These are held in place at both ends by brackets that run down each side, bolted to the flat wooden base. The track rods are just lengths of ½" rigid steel electric conduit tubing. The brackets are L-shaped aluminum bars, which I drilled at both ends so I could slide the tracks through them.

The carriage track, which carries the plotter's stylus, is oriented up-and-down to control the y-axis. The entire carriage slides left and right on top of the main track, which supports it. After a bit of experimentation, I realized that I needed some additional clearance between the carriage and the baseboard. So, I raised the main track off of the board by adding some nylon washers onto the bolts underneath the aluminum brackets.

The carriage has two structural pieces: the frame and the track. The frame is a length of 1" square aluminum tubing, which bears the weight and carries the motor, pulley, and track assembly. The track is a smooth ¼" aluminum rod running parallel to the carriage frame that the stylus bearing slides along.

To get the carriage frame to slide along the main tracks, I attached two ½" internal-diameter linear bearings, which I bought from Skycraft Parts and Surplus (skycraftsurplus.com). These fit around each rod like cuffs, and they slid up and down the tracks perfectly, but mounting them to the square-shaped box tubing was a bit of a challenge. I eventually cut the ends of each tube down 1" on 3 sides and glued the bearings into the cutouts with JB Weld epoxy. I had to make sure that the bearings on each side were perfectly parallel, so they would not bind when trying to slide along the main tracks.

I made two brackets to attach the ¼" carriage track rod to the square tubing of the carriage frame and slipped on a matching internal-diameter linear bearing from Skycraft to hold the stylus. With the carriage completely assembled and sliding along the main track, the Scribbler Bot's framework was complete.

Attaching the Motors and Pulleys

The stepper motors use pulleys and belts to move the carriage and stylus into position. Getting the carriage to move along the main tracks (x-axis) is the heavier problem; the tracks are far enough apart that the carriage needs to be pulled along both sides. I decided to use an axle to connect two pulleys on the side opposite the motor. The force would transmit in a "C" shape starting from the motor: motor to belt, to pulley, to axle, to opposite pulley, to opposite belt.

I used a ¼" automotive timing belt for the belts on both sides. For the first three points in the drivetrain, the motor and the two pulleys on the axle, I attached sprockets to mesh with the belts; the last point in the drivetrain is just a smooth drum. I attached all 3 pulleys to spin on bearings from an old skateboard. For the axle, I used more ¼" aluminum rod.

I mounted the x-axis motor and the last pulley onto the baseboard using some shelving brackets and random bolts. For the pulleys on the axle, on the opposite end of the main tracks, I fabricated custom brackets to attach them to the board. The belts travel directly over the main tracks and attach to the carriage frame underneath, in order to pull it along. To make this connection, I drilled holes through the underside of each belt and the top of the carriage frame, and secured them together with small bolts.

The carriage uses just 1 belt to move the stylus. I mounted the y-axis motor on one side of the carriage and a pulley on the other. I attached a sprocket onto the motor shaft using my plastic-straw trick (described next) and fitted more ¼" timing belt around the motor and pulley.

With both motors, I ran into a problem attaching the sprockets to the motor shafts. I couldn't find sprockets that matched the belts and had the right internal diameter to fit over the motor shaft. I looked at the ordinary plastic straw in my soda and realized it was the same diameter as the inside diameter of the sprocket, so I stuck it into the sprocket, cut it off, and filled it with JB Weld. Once it cured, I drilled a new, smaller hole in the center, which fit over the motor shaft perfectly.

Finally, it was time to hook up the controller board and start testing. I soldered motor wires to terminals on the controller board and, after some trial and error, got everything to work. Since then, the Scribbler Bot has drawn dozens of portraits, including many at last year's TED conference in Monterey, Calif. (ted.com). Meanwhile, I have improved things by putting all of its electronics into a detachable project box that the control board and motors can easily plug into.

Future Directions

The Scribbler projects are constantly evolving. I am currently working on a web version that will allow users to scribble images and watch them be drawn on the Scribbler Bot via webcam online. Also in the works is a wireless sidewalk-chalk bot that will drive around and draw images based on input from users over the internet.

The Scribbler

The Scribbler Bot's software derives from an online drawing toy called the Scribbler (zefrank.com/scribbler), written in Flash by Ze Frank. Since the Scribbler, Ze and I have collaborated on other projects based on its algorithms, which generate complex line drawings out of any image.

With Scribbler Bot, we wanted a machine that creates keepsake drawings of people's faces quickly and recognizably, while maintaining a sense of originality and surprise. "Analog distortions" from motor vibrations, broken pencil lead, and other causes are part of the fun. For the Bot version, I ported the original Flash code to Director, for better speed, motor control, and camera integration.

The Scribbler was influenced by the artbot scene (artbots.org). For more information on the Scribbler project, please visit lotsofwires.com.

Douglas McDonald is a multimedia developer from Orlando, Fla., who enjoys working on art robotics in his spare time. For more information on his projects, please visit tek-tonic.com.

Me and Judge Alito, at his confirmation hearing. I'm sitting at far left, enjoying a cold brau.

HOW TO DRINK BEER ON C-SPAN

Put yourself into somebody else's video.

By Bill Barminski

OK, you're not really going to drink beer on C-SPAN or *Larry King Live*. But you can make it look like you did on video. I don't know why you'd want to, but let's just say you do. I know I did.

The method used to achieve this effect is called *compositing*. You will need a source video recorded from a television show, a replacement video you will shoot yourself, and a *static matte* — a shape cut out of the source video with Photoshop to hold the new video.

The first step is to watch TV and record the source video. Sounds like fun, right? Don't get too excited. I recommend using video from C-SPAN, which is a good source for two reasons. First, they repeat everything, over and over and over. So if you see something good, you can catch it later and record it. I use an analog-to-DV converter on my computer to capture stuff live off the TV directly into Final Cut Pro. But you can record onto a DV camera and then later capture it into the editing software of your choice.

The second and more important reason why C-SPAN is a good source is that the network tends to use "locked-down" camera shots. (A *lockdown* is a camera that has been set on a tripod and is not moving or panning.) No matter where you get your source video, you need to look for shows that use locked-down camera angles. A locked-down

Photography by Bill Barminski

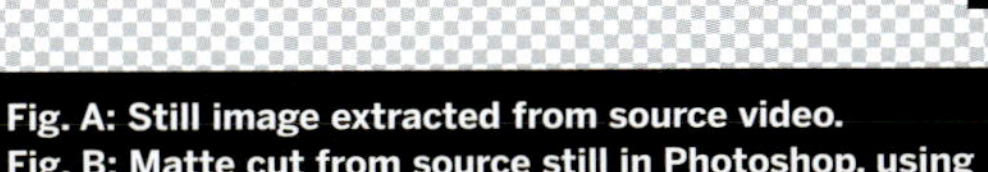

Fig. A: Still image extracted from source video. Fig. B: Matte cut from source still in Photoshop, using the Path tool. Fig. C: The checkered area is transparent. Fig. D: Shooting the fake video.

camera dramatically reduces the amount of work required. If the camera is moving, the shape of the matte you create has to move, too, in every single frame of the video captured, which is 30 frames per second! Do the math. A 10-second shot will require you to cut something like 8 billion mattes. OK, my math may be off, but it's still a lot. With a locked camera shot, you just need one, which you can cut in Photoshop.

MATERIALS
C-SPAN
Analog-to-video capture device
Final Cut Pro and Photoshop
Video camera
Tripod
Beer
Microphone to record belch
Business suit

I recorded some footage of Judge Alito at his confirmation hearings. In the shot I used, the camera is locked down and Judge Alito is sitting mostly still. More importantly, the woman wearing light pink in the row in back is very still. I want to sit next to her and drink beer.

Once I've found a source video clip, I extract a still image from it, open it in Photoshop, and cut the matte. I use the Path tool to make this selection; a 1-pixel feather on the selection is recommended. I place the selection in a new layer and fill it with any color. I turn off the background image, leaving me with a solid shape. I save this as my matte with the alpha/transparency intact. From the File menu, do a Save for Web with PNG-24 selected as the format and the Transparency option checked. The checkered background indicates the transparent area.

Once the matte is cut, it's a good idea to test it to see whether it really holds up before wasting time shooting a replacement video. Using Final Cut Pro, I place my original video into Layer 1. Then I import my PNG matte and place it into a new layer above the original video. It should fit right into place. I render the shot and check to see whether the edge along the woman in pink's silhouette will work. If she sits still and the camera doesn't move, then the matte should sit perfectly along her outline. If she picks up her hand, it ruins this matte because her hand would travel into the orange area and get cut off.

Here I am again, seeing if the Alito hearing is any more interesting if I'm sitting behind Ted Kennedy.

Now Comes the Fun Part: Making Your Fake Video

I put on the only suit I own and set up my camera on a tripod in my living room. In order to get the camera angle right, I print out a still image from the Alito footage and make adjustments until I have the same approximate angle. I don't worry about placing myself in the corner of the shot; in fact, I shoot myself central to the frame, knowing I can hand-place the shot in the corner later.

I originally shot myself doing various things such as drinking beer and making faces and hand gestures. But I found that the faces and hand gestures were just too obvious, too over the top. The beer drinking was subtle. In fact, when I showed the finished composite shot to some people, they had to watch it twice before figuring out that something wasn't right. I consider that success.

Once the fake video is shot, I'm ready to composite all the elements together. I use Final Cut Pro, but After Effects or Premiere will also work. I import my three elements — the original video, a PNG matte, and the replacement video — into a project. I place the original video into the bottom video layer, the PNG matte into a new video layer above it, and the fake video in a third layer above the first two.

The method used to put the fake video into the matte area of the real video is called *travel matte alpha*. I double-click to select the fake clip, open the Modify column, and then, near the bottom, open the category Composite Mode. I then select the Travel Matte Alpha option near the bottom of that subcategory. The fake video now plays only inside of the shape of the matte.

The video may now need to be moved into position to look correct. I had to resize mine a little and move it to the left until I was "sitting" next to the girl in pink. I also did some color correction so that both shots looked consistent.

The last step was to record a loud burp. I used the voice-over option in Final Cut to record that additional sound. I'm done, unless I want to eat chips behind Senator Kennedy.

These clips can be seen at barminski.com.

Bill Barminski is a multimedia artist currently teaching in the School of Theater, Film and Television at UCLA.

MIDI menagerie: Mini blinking monster pals help musicians debug their hardware setups.

MONSTER MIDI DETECTOR

Here's an easy-to-build MIDI data detector packaged in a small Japanese action figure.

By David Battino

Photography and illustration by David Battino

For a technology invented 23 years ago, the Musical Instrument Digital Interface (MIDI) is amazingly flexible. This underappreciated communications protocol now graces gadgets from cellphones to kilobuck professional keyboards; it even powers theme park rides and Las Vegas shows.

The original concept behind MIDI was to enable musicians to layer the sounds of multiple hardware synthesizers. Connect a cable to the MIDI Out jack of one keyboard and the MIDI In jack of a second, press a key on the master keyboard, and both instruments will sound. That is, if they're configured correctly.

If they're not, you'll run into problems. For example, if the master device is transmitting on MIDI channel 3 and the slave is set to receive on channel 4 (there are 16 MIDI channels), you won't hear anything.

Similarly, most master keyboards can generate additional performance information beyond just the notes played, such as aftertouch or control change data — for example, when you press harder on a key that's already down or move a wheel to bend a pitch. But if you haven't configured your MIDI settings correctly, those transmissions may be disabled in the master or ignored by the slave.

Software MIDI monitors offer more detail, but for testing, a hardware detector is unbeatable. This lets you debug your setup by confirming that

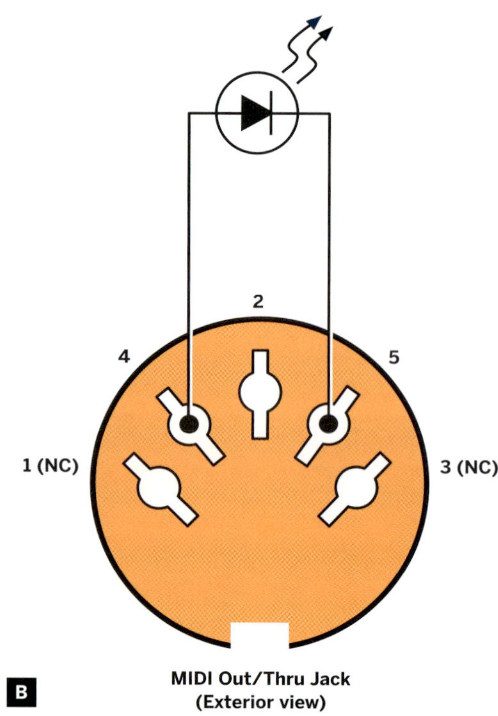

A An LED between pins 4 and 5 on a MIDI Out or Thru jack will flash when data passes. Fig. A: When wiring a receptacle remember that the pins are reversed. You can also connect directly to a captive MIDI cable.

B Fig. B: You can wire a 220-ohm resistor in series with the LED to protect it, but for quick checks, it's perfectly safe without one.

MATERIALS
- High-intensity LED
- 5-pin DIN jack or MIDI cable
- Insulated wire or wire wrap
- Hollow toy
- (Optional: 220Ω resistor, LED socket)

TOOLS
- Soldering iron or wirewrap tool
- Drill
- Tapered reamer
- Long-nose pliers
- Wire cutters/stripper

your instrument is generating MIDI messages for both raw notes and any additional twiddling. If the detector shows that the MIDI is coming out of your instrument, you know you have to look elsewhere for the reason you're not hearing anything.

To determine if my keyboard was successfully generating data, I built this simple MIDI detector and packaged it in a small Japanese monster toy. Besides looking cool, the toy — a Bandai Mechanikong — pulled apart easily to grant access to its insides. It also had a flat spot on the back where I mounted the MIDI jack. I've since built three more monster MIDI detectors (see previous page).

How It Works

Only the three center pins in a MIDI Out jack are connected. Pin 2 is ground, pin 4 is +5VDC, and pin 5 is normally at +5VDC, but it drops to 0VDC when MIDI data goes out. So if you connect the anode (long leg) of an LED to pin 4 and the cathode to pin 5, the LED will flash when MIDI data goes out (see Figure B). Note that MIDI pins are numbered non-consecutively, with pin 2 in the middle, between pins 4 and 5.

You can also use this device as a visual metronome. Just send it a control-change message every quarter note, with a burst of messages to mark the downbeat. (If sending from a sequencer, you may need to filter out MIDI Clock data, which pulses 24 times per quarter note.)

See a movie of the Mechanikong MIDI detector in action at makezine.com/07/diy-monstermidi.

David Battino (batmosphere.com) edits the O'Reilly Digital Media site (digitalmedia.oreilly.com) and co-wrote *The Art of Digital Music*.

Dig-it: One microcontroller pin connects to each segment of a seven-segment LED.

HELLO, WORLD

Programming microcontrollers, part 2.

By Sparkle Labs

In part 1 of our microcontroller programming primer (*MAKE, Volume 04, page 158, "Microcontroller Programming"*), we explained how to get started using microcontrollers. We put together a development environment and built a battery-powered LED blinky that's controlled by a PIC chip. The blinky's on/off timing was determined by a simple BASIC program, which we compiled using PICBasic Pro within the MicroCode Studio development environment and uploaded to the chip using a PICkit 1 programmer board.

This article continues up the micro project food chain, showing you how to make projects plug-powered, how to read buttons and variable-resistance components as input devices, and how to program seven-segment LEDs and alphanumeric LCD display modules as output devices.

Photograph by Sparkle Labs

Plugging In

We used batteries to power our LED blinky in part 1. This is the simplest solution, but batteries wear down, and plug-in power is usually better for indoor applications. For our next projects, we'll use a standard wall-wart AC-DC transformer and connect a 7805 power regulator to protect the circuits from too-high voltages. This arrangement can replace a 5V battery source in any circuit.

The PIC Chip

We used Microchip's eight-pin PIC 12F675 microprocessor for our LED blinky in part 1, but we need more pins in order to address multisegment LED and LCD devices. Here, we'll use a PIC 16F684, which has 14 pins and also fits into and works with our little PICkit 1 programmer board.

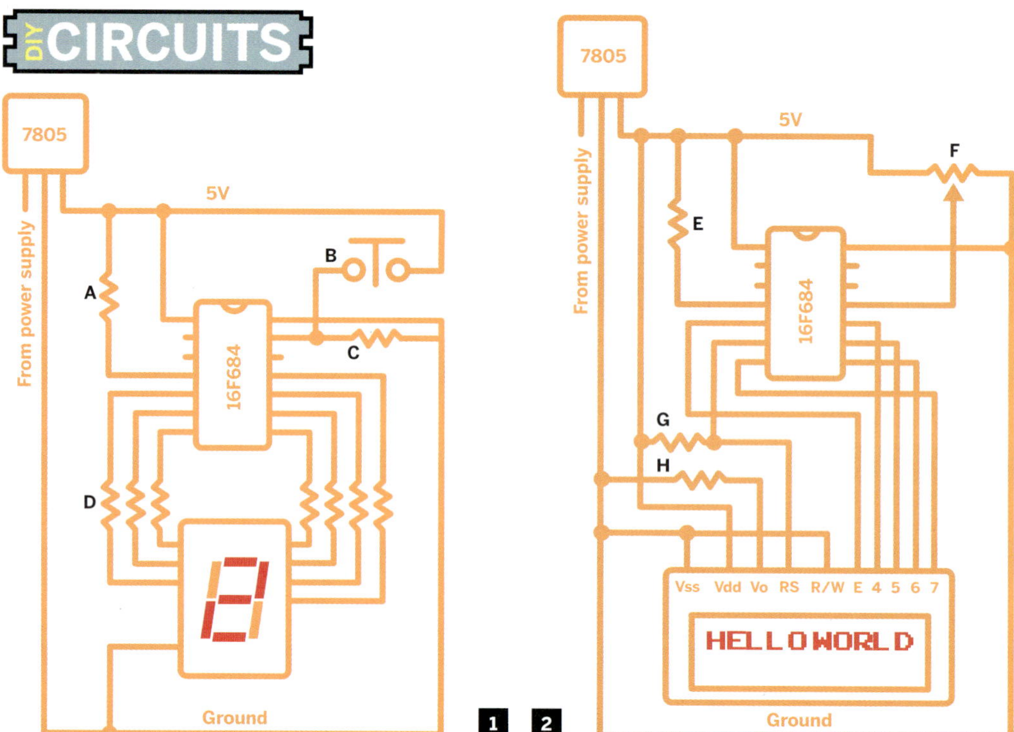

Fig. 1: LED button-press counter circuit: A. 10Ω resistor; B. SPST button; C. 10kΩ resistor; D. 220Ω resistors. Power supply wiring at upper left provides constant voltage from a plug-in transformer.

Fig. 2: Potentiometer knob with LCD circuit (shown programmed for static display test); E. 10Ω resistor; F. 10kΩ potentiometer; G. 10kΩ resistor; H. 1kΩ resistor.

Project 1: LED Button-Press Counter

First, let's put together a circuit that shows the number of times you push a button. For the display, we'll use a seven-segment LED, which consists of seven LEDs arranged to show a single digit. The package has eight pins: a power pin for each of the segments and a shared ground. To hook it up, we wire the ground pin to ground, and connect each LED anode to its own pin on the PIC, which we set as output in our software. This way, each segment becomes addressable as a different pin (or port, in the code) on the chip.

For the button, we designate one pin on the PIC as a digital input. Referring to the chip's datasheet, we set registers at the top of our program to tell our PIC to use pin 13 as a digital input and pins 5–11 as digital outputs. (See code online.)

Subroutines for each displayable numeral define which LEDs are on and which are off. For example, the display3 routine shows a "3" by turning on all the segments, except for the top-right and bottom-right ones.

To wire up the button (or "momentary switch"), you connect the input pin to one side of the button and also to ground (0V) through a resistor. The other side of the button connects to power (5V). When the button isn't pressed, the input reads low. In code, this means port A0 (pin 13) equals 0. Pressing the button brings power to the input, making port A0 equal 1 in the code.

We keep track of the count by defining a variable, countVar, which increments (or resets to zero) each time the button is pressed. The main routine loops forever, reading the input pin, changing countVar, and calling the digit display subroutine.

Our main program loop might go through countless iterations during each button press. We only want to increment countVar the first time we read a press, not every time. So our code uses the variable buttonState to track whether the button was just pressed, or was in that state already.

In practice, buttons don't always form a solid immediate connection; sometimes there's a tiny stutter as they make contact. To handle this, our code pauses for ten milliseconds after reading a button-pressed state, to let the button settle:

```
countVar var byte 'Keep track of the count.
buttonState var byte 'Keep track of the button state.
buttonState = 0 'State 0 means the button is ready.
```

Illustrations by Sparkle Labs

```
main:
  if portA.0 = 1 then 'If the button is pressed then...
     pause 10 'Let button settle to prevent false readings.
  if buttonState = 0
        countVar = countVar + 1 'Increment countVar by 1.
        if countVar > 2 then 'When it goes above 2 return to 0.
           countVar = 0
        endif
        buttonState = 1 'Button is not ready.
     endif
  else
     buttonState = 0
  endif
  if countVar > 2 then
     countVar = 0
  endif
  gosub displayNum 'Call subroutine to display the count.
goto main
displayNum:
  branch countVar,[display1, display2, display3]
return
```

Project 2: Potentiometer with LCD Readout

LCD modules are a simple way to display alphanumeric characters. Unlike with the seven-segment LED, you don't have to program each bit of each character to create readable text. The PicBASIC Pro compiler offers the handy BASIC command LCDOUT, which takes a text string and pushes it onto a compatible display. To use it in your code, you simply specify which pins are connected to the LCD module, and then pass strings to LCDOUT to display them.

Full documentation on using LCDOUT, including LCD module wiring, is in the PICBasic Pro Compiler Manual on microEngineering Labs' website (melabs.com/resources/pbpmanual/).

Following the pin-out diagram of the LCD, we wire it to available pins on the PIC. We connect the LCD's power and ground, and ground the contacts that are for features we aren't using.

Now we can program the micro to test the LCD by showing the traditional "Hello World" greeting. After declarations at the top, our main loop contains just two lines. (See full code online.)

```
main:
  LCDOUT $fe, 1 , "Hello world."
  pause 1000
goto main
```

Unlike a button, a potentiometer is an analog input device. Turning its knob changes its electrical resistance. Many light sensors, proximity sensors, flex sensors, and other types of input devices also work as variable resistors. They're all wired into a circuit and handled programmatically in the same basic way, so learning how to read these with a microprocessor opens up countless possibilities for inputs and interactivity.

We'll program our circuit to read the value of the potentiometer and display a corresponding number, between 0 and 255, on the LCD. We wire one pole of the potentiometer to power (5V), the other pole to ground (0V), and the variable "wiper" to pin 11. In the code's declarations, we set pin 11 to be an input pin and enable it for analog digital conversion (ADC). This means the chip will read variable voltage in the 0–5V range as a number between 0 and 255. Turn the potentiometer knob, and you change the number displayed.

In practice, the voltages read by the microprocessor are noisy, and immediately pushing raw readings to the display makes the numbers jumpy and unreadable. Our code solves this problem by tracking the previous display reading and updating it only if there's a significant change to the input, with the threshold set at 7.

```
adcVar var byte 'Analog reading value.
prevadcVar var byte 'Previous reading.
main:
adcin 2, adcVar 'Read the ADC value of analog pin 2.
if (ABS (adcVar - prevadcVar)) > 7 then 'If significant change...
  LCDout $fe, 1, "adcVar =" 'Clear screen and print "adcVar =".
  LCDout $fe, $C0, #adcVar 'Print value itself on 2nd line.
  prevadcVar = adcVar
endif
goto main
```

Now you can use all sorts of input devices and displays! Enjoy, and please join us for part 3, in which we'll be making a digital alarm clock.

+ Materials list, supplementary code, additional diagrams, and suggested reading online at makezine.com/07/diycircuits_hello.

Sparkle Labs (sparklelabs.com) is a product development firm in New York City. They build "hi-tech, hi-touch" environments and products, using new technologies.

MAKER TRIPS Duct Tape and Cyclotrons

By Michael Shapiro

Incredible machines at the Lawrence Berkeley National Laboratory.

"There's a little lab down there," said Paul Preuss, a guide at the Lawrence Berkeley National Laboratory, "where they discovered that duct tape is good for just about everything except sealing ducts." He wasn't kidding — lab scientists really ran an experiment.

"They tried all sorts of materials: clear plastic tape, foil-backed tape, aerosol sealant; and everything worked well, except duct tape. It would just dry up and fall off," Preuss said. "But it wasn't meant for sealing ducts — it was created for temporary battlefield repairs during World War II."

I'd come to the Berkeley Lab (www.lbl.gov) hoping for a glimpse of machines that propel electrons to almost light speed, and here I was, standing beside a grove of Chinese dawn redwoods with a spectacular view of the Berkeley campus and the Golden Gate Bridge, learning about the shortcomings of duct tape.

But that was just the beginning of the tour. During the next couple of hours, Preuss led me around the lab's 183-acre hillside campus, introduced me to scientists working at the cutting edge of electron microscopy, and took me inside the synchrotron at the Advanced Light Source (ALS), where a ring of pipes, energy boosters, and magnets shoots electrons to such high speeds that they become relativistic. In other words, as Einstein postulated a century ago, the subatomic particles take on greater mass as they approach light speed.

Founded in 1931, the lab was the brainchild of Ernest Orlando Lawrence, the physicist who won the Nobel Prize in 1939 for his invention of the cyclotron, a circular particle accelerator that opened the door to high-energy physics. Since then, Berkeley Lab scientists have discovered 16 chemical elements and won 10 Nobel Prizes.

The first cyclotron was five inches in diameter and contained two D-shaped semicircles. Protons were injected, accelerated by magnets outside the cyclotron. The particles accelerated each time they passed between the two halves of the cyclotron, spiraling to higher and higher speeds. Next, an 11-inch cyclotron was built, and just before the United States entered World War II, construction began on a gargantuan 184-inch cyclotron.

That may not sound very big, but with the surrounding pipes and 4,500 tons of magnets, the beast took up as much space as a baseball diamond. The project was put on hold during the war but was completed in 1946. The new "synchrotron" at the lab's Advanced Light Source was built around the old cyclotron; you can still see where the old walls were.

Enclosed by a three-foot-thick wall of concrete and lead (to prevent radiation from leaking out), the synchrotron is off-limits while it's running and, thus, can't usually be observed. Because the synchrotron was down for routine maintenance the day I visited, Michael Martin, one of ALS' leading scientists, offered to show us around. The ALS, he explained, generates intense beams of light for scientific research that ranges from breaking down toxic waste with basalt-eating bacteria to evaluating the particles in a comet's tail. The lab began doing the latter after the *Stardust* spacecraft returned to Earth in January with samples from the Wild-2 comet.

Since 1939, Berkeley Lab scientists have discovered 16 chemical elements and won 10 Nobel Prizes.

Martin led us through the door — also three feet thick, it looked like a bank vault door on steroids — into the heart of the ALS: the synchrotron. The ring accelerator has 12 sections, each of which is controlled by three mammoth magnets. The blue ones, with a strength of approximately 1.2 tesla, are the bend magnets, Martin explained, which help accelerate and focus the particles inside. Vat-like copper "kettles" use electric fields to accelerate "tired" electrons. "What makes the electrons give off light (from infrared radiation through X-rays) is that they are forced to turn corners or wiggle," Preuss said. "The faster they go, the more energy they give off when they turn or wiggle."

The makers of the synchrotron came up with some cool techniques to keep everything in sync. Each magnet around the central pipe has six struts for adjustments up and down, forward and back, and left and right. This precision enables the beams of light to maintain accuracy within a few microns. About three dozen pipes and sub-pipes called "beamlines" shoot off the main ring. Some beam-

Photograph by Michael Shapiro

The synchrotron ring accelerator at the Berkeley Lab. Accelerating particles almost to light speed, the synchrotron is off-limits while it's running and can't usually be observed.

To visit the Lawrence Berkeley Lab: Located in the Berkeley hills above campus, LBL conducts several tours each month. Tours start at 10 a.m. at the Berkeley BART station and last about 2½ hours. To request a tour, visit www.lbl.gov/Community/tours.html or call (510) 486-7292. Reservations should be made at least two weeks in advance. Typically, a tour includes visits to two or three research areas. Photography is allowed in most labs, and wearing comfortable footwear is advised.

Tour participants may be asked for photo ID and citizenship information. The Berkeley Lab is not associated with the Lawrence Livermore Lab and conducts no classified or weapons research.

lines have been used for X-ray microscopy; one early achievement was viewing the malaria parasite inside a red blood cell. Others are used in nanostudies and for superconductors, aiding research in microelectronics.

The tour moved on to the National Center for Electron Microscopy (Building 72), which houses the rocket-like Atomic Resolution Microscope — the most powerful electron microscope in the world when it was unveiled in the mid-80s. It is "the machine around which this place was built," Preuss said.

We dropped in on Erdmann Spiecker, a visiting scientist from Germany's University of Kiel. Using an ultra-powerful scanning electron microscope, Kiel discovered a new way to form complex networks of nanotubes by depositing metals on the surface of layered crystals. The Berkeley Lab proved that what were once thought to be cracks are actually a network of hexagonal nanotubes that form when enough metal is absorbed by the crystal's surface layers.

Before leaving Building 72 we visited Andreas Schmid, who works with a low-energy electron microscope, or LEEM. Other electron microscopes focus 100,000 volts to such a small density "that it's surprising the sample doesn't just disappear," Schmid said. "The power density is 10 times higher than the 192 lasers used for nuclear fusion!" But the LEEM is different; it hits the sample with a trillion times less energy. It's used to record images of the top few atomic layers, ideal for exploring the fabrication of computer chips and circuits. "If you want to make a better hard disk," Schmid said, "just come here."

Michael Shapiro is the author of *A Sense of Place: Great Travel Writers Talk About Their Craft, Lives, and Inspiration.*

BY SAUL GRIFFITH, NICK DRAGOTTA & JOOST BONSEN '06

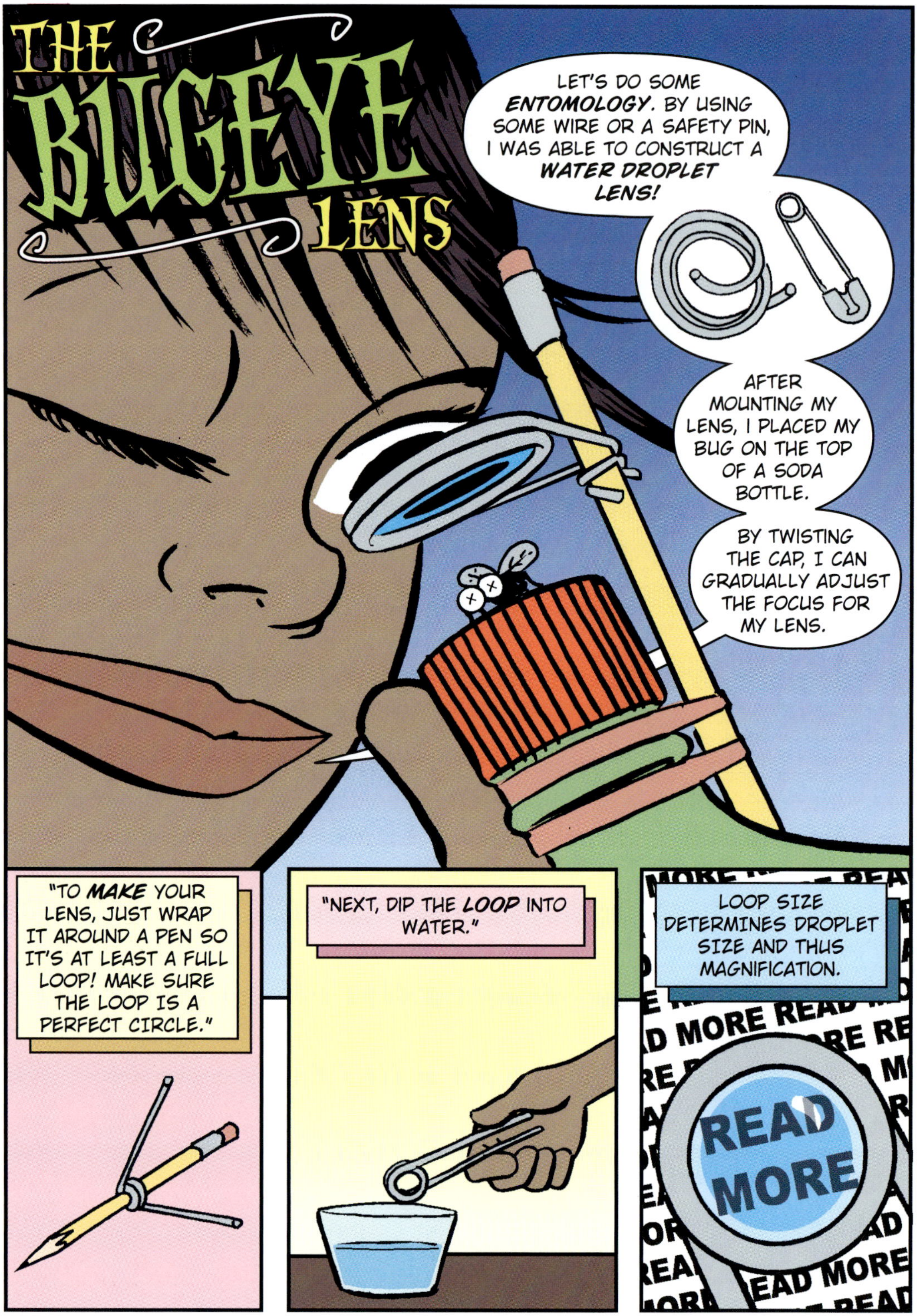

THE BUGEYE LENS

LET'S DO SOME **ENTOMOLOGY**. BY USING SOME WIRE OR A SAFETY PIN, I WAS ABLE TO CONSTRUCT A **WATER DROPLET LENS!**

AFTER MOUNTING MY LENS, I PLACED MY BUG ON THE TOP OF A SODA BOTTLE.

BY TWISTING THE CAP, I CAN GRADUALLY ADJUST THE FOCUS FOR MY LENS.

"TO **MAKE** YOUR LENS, JUST WRAP IT AROUND A PEN SO IT'S AT LEAST A FULL LOOP! MAKE SURE THE LOOP IS A PERFECT CIRCLE."

"NEXT, DIP THE **LOOP** INTO WATER."

LOOP SIZE DETERMINES DROPLET SIZE AND THUS MAGNIFICATION.

READ MORE

From the forthcoming book *HOWTOONS Book 1* by Saul Griffith, Nick Dragotta, and Joost Bonsen. Published by arrangement with ReganBooks, an imprint of HarperCollins Publishers, Inc.

MakeShift

By Lee D. Zlotoff

The Scenario: You set off on a solo backpacking jaunt one blissfully free weekend, in search of a legendary mountain hot spring that has remained pristine thanks to the 12-plus-hour climb it takes to reach it. A well-earned sweat topped off with nothing but silence, solitude, and hot water — what's not to like?

Just as your topo map indicates that you're within minutes of the spring, you hear an agonized shouting from somewhere off the rocky trail. You quickly discover a large, cylindrical fissure in the ground, about 15 feet in diameter and about 20 feet deep, at the bottom of which lies a rather large example of humanity, with his leg bent at such an unnatural angle that there's no doubt it's badly broken. You yell down to the man — who is easily twice your weight — to say help has arrived. He acknowledges you with a wave, but he seems to be fading fast from shock, or pain, or whatever. The walls of the fissure are nearly vertical and full of jagged rocks, but your experience tells you they're scalable. Still, there's no way you'll be able to climb those walls with this guy on your back. You'll have to come up with another way to get him out of this hole.

And then it hits you: A noxious, sulfuric smell that says that this fissure is a vent for the same gases that make the hot springs so warm and bubbly. If you don't quickly find a way to get fresh air to this guy, he's not going to survive *long enough* for you to rescue him.

The Challenge: Devise a way to keep this guy breathing while you come up with and execute a plan to safely extract him from the fissure. Then get him stabilized enough that you can either get him off the mountain yourself, or hike back out to summon more help.

Here's what you've got: A top-of-the-line backpack with a nested, detachable water container, a sleeping bag, inflatable air mattress, two-man backpacking tent, a large towel, cook set, butane stove, camping food, and a basic first aid kit. You also have about 40 feet of nylon rope, an elaborate Swiss Army knife (or Leatherman tool), a 25-foot roll of duct tape, a small Maglite-type flashlight, your trusty, 6-foot bamboo walking stick, and the bandana around your neck. Any questions? Good, 'cause humanity awaits.

Send a detailed description of your MakeShift solution with sketches and/or photos to *makeshift@makezine.com* by Oct. 27, 2006. If duplicate solutions are submitted, the winner will be determined by the quality of the explanation and presentation. The most plausible and most creative solutions will each win a MAKE sweatshirt. Think positive and include your shirt size and contact information with your description. Good luck! For readers' solutions to previous MakeShift challenges, visit *makezine.com/makeshift*.

Lee D. Zlotoff is a writer/producer/director among whose numerous credits is creator of *MacGyver*. He is also president of Custom Image Concepts.

Photograph by Christopher Lucas

ModBots, by Bil Bowen and Eric Singer, are modular robots designed to attach to any structure, from stairwells to Tibetan bowls to ship hulls. MIDI messages are converted electrically to mechanical motion via custom circuits.

Photography courtesy of LEMUR

MIDI Control

Music equipment language isn't just for audio anymore.

By Peter Kirn

For centuries, musical notation has served as a common language among musicians — it was designed so that, for example, monks in France would sing the same melody as monks in Rome. But as the popularity of digital musical instruments grew throughout the 1980s, musicians found that their equipment lacked a common language.

There was no way to perform simple tasks like using one keyboard to play sounds on another, or to use a computer to record and edit what you were playing. The Musical Instrument Digital Interface (MIDI) was developed as a solution to this problem, and today, it's become a standard for the vast majority of music hardware and software. Its usefulness doesn't end there, however. The same structure that makes MIDI compatible with various music products can make it useful any time you need to send and receive messages for control.

An Introduction to MIDI

MIDI doesn't actually represent sound at all; instead, it's a standard specification for storing and transmitting the events that produce sound, like strumming a guitar or pressing keys on a keyboard. In this way, MIDI is very much like sheet music — it specifies what someone plays, not the resulting sound. Because MIDI is a digital format, the information is numeric and can be generated by and used to control anything. As soon as MIDI was introduced, in the early 1980s, electronic musicians started to experiment with it — if the software expects MIDI messages for specific pitches, such as from a standard piano keyboard, why not feed it from drum triggers so you can play keyboard licks on your drums? Why not take it further and play musical notes from a video game controller, a Wacom graphics tablet, or even sensors embedded in your pants? (Yes, that last example is real — players have been known to put drum triggers on their legs.)

Because it's ultimately just numbers that represent events, MIDI isn't even limited to music. MIDI and extensions to the MIDI protocol have been used to control theatrical lighting, fireworks, robotics, flame-throwing organs, and even the visual effects of Disneyland-style theme park rides. This article looks at how you can use MIDI to control your own projects, whether or not they generate music.

The best part about MIDI is its compatibility. You have off-the-shelf and DIY options that speak MIDI on both the hardware and software sides (MIDI instruments and controllers; hardware interfaces; and music, video, and multimedia software). You'll even find MIDI support in Final Cut Pro, which means you could edit your next film on a music keyboard.

So, you could build a custom hardware controller for your video software and plug it into your computer without writing drivers or coding. And if doing this lets you focus on the design and have fun with your invention, it's clearly worth it!

Jacks and Cables

The standard MIDI connector is a male serial DIN plug. It has 5 pins, but only the 3 center pins carry voltage. Data flows through MIDI in only one direction at 31,250 bits per second. Because 5-pin DINs aren't exactly standard equipment on computers, use a MIDI jack included on an audio interface or sound card, or use an adapter. For $40 you can buy a compact 1-in/1-out MIDI-to-USB interface, such as the M-Audio Uno (maudio.net) or Edirol UM-1EX (edirol.com). Interfaces with more jacks are available for more complex setups.

MIDI cables are unidirectional, which dictates how your equipment is connected. If you want one MIDI device to transmit signals to another, connect the "out" jack of the transmitting device to the "in" jack of the receiving device. If both devices need to send data to each other, as you might do with a MIDI sensor board and a computer, you'll need two cables.

Some MIDI hardware aimed at computer users, such as the M-Audio Oxygen8 keyboard, has USB or FireWire jacks for MIDI output in addition to the standard MIDI port. This lets you plug the hardware right into a laptop without an adapter.

Understanding MIDI Message Types

There are three basic kinds of MIDI messages:

1. Notes — On a keyboard, you press a key to generate a note-on event, accompanied by a number for the pitch and a velocity number describing how hard it was struck.

The note of that pitch will continue to sound until an instrument receives a note-off event. In non-music applications, note messages can trigger specific events, such as lighting or video cues.

2. Controller messages add expressiveness by encoding the changes in volume, tone, and pitch. For non-music applications, controller messages work well for values that change over time, such as a temperature sensor level or a knob position.

3. System messages provide sync and timing, system real-time messages, and MIDI Show Control (MSC) extensions that allow use with applications, such as lighting.

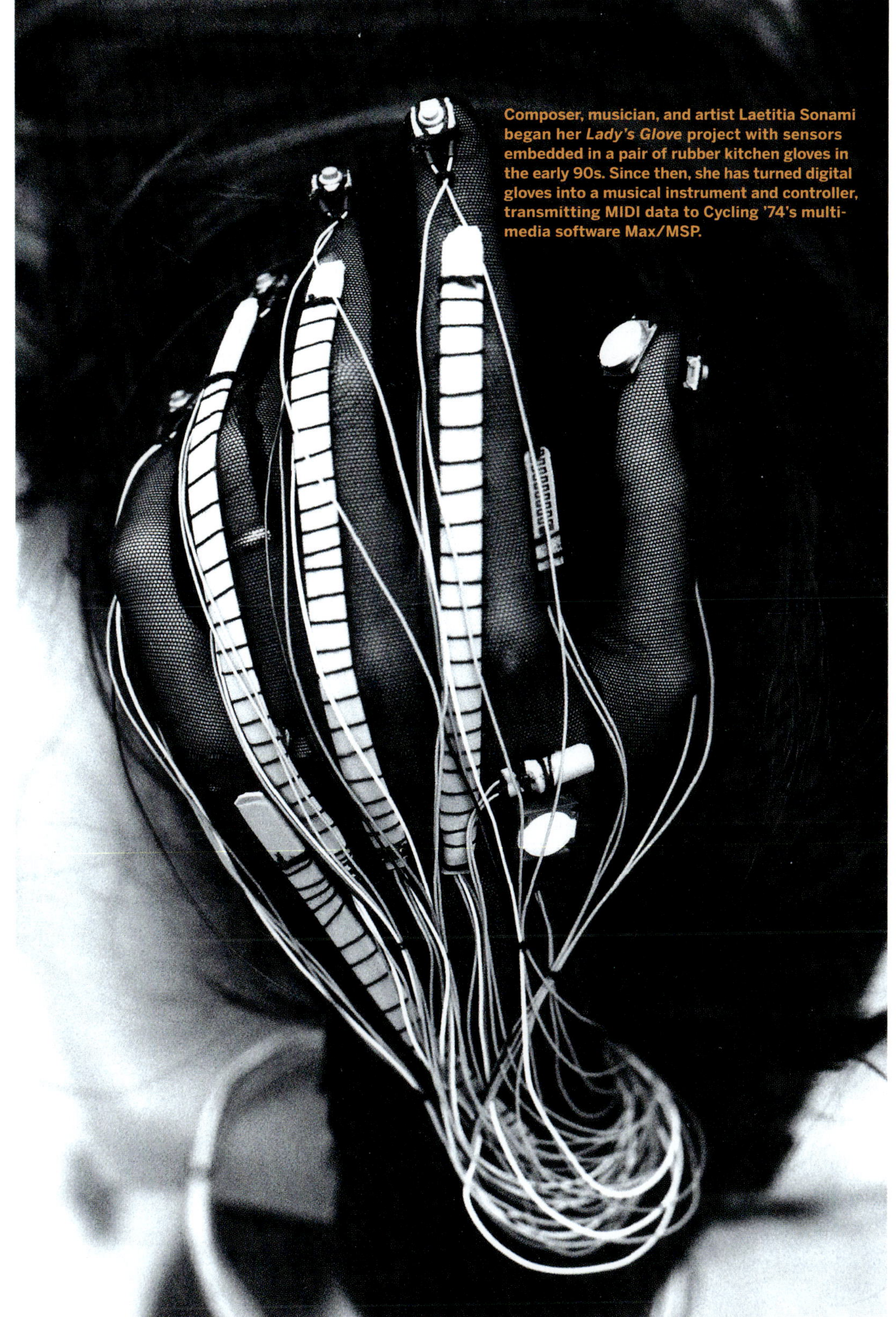

Composer, musician, and artist Laetitia Sonami began her *Lady's Glove* project with sensors embedded in a pair of rubber kitchen gloves in the early 90s. Since then, she has turned digital gloves into a musical instrument and controller, transmitting MIDI data to Cycling '74's multimedia software Max/MSP.

Photograph by A. Hoekzema

Using MIDI Message Types Creatively

Here are some tips for designing a control setup to take advantage of the way the MIDI protocol is structured.

Route your data using channels. MIDI lets you connect several devices together in a chain and use them all at once. All the chained devices receive all of the messages, so MIDI uses channels (1-16 for each MIDI port) to point certain messages at certain devices. Transmit data on channel 3 and only those devices that you've set to receive on channel 3 will respond. This is an easy way to direct data to and from different parts of your system without worrying about how to wire them together. Just connect all your devices in one chain and set their input and output channels to route the messages along the paths you want.

Pick your controls. While channels can direct data, control messages can describe data. If you have 8 sensors or 8 controls, assign each a different Control Change (CC). You can choose from MIDI's undefined CC numbers or use defined values, such as 7, which denotes volume. In a nonmusical application, a volume control could translate to the brightness level of an LED.

Scale the data appropriately. Contrary to popular belief, MIDI isn't limited to 128 discrete values. For example, the pitch bend control (which is unique and doesn't have a CC number) uses about 16,000 data levels. Other CC controllers achieve greater resolution by combining two bytes of data, for coarse and fine levels. (Programmers who have worked with Most Significant and Least Significant Bytes (MSB/LSB) are familiar with this technique.)

This is another factor to consider when you pick which MIDI controls you want to assign to the sensors or other inputs in your project — you might not need pitch bend, and a wider range of values isn't always better. Resolution that is too high requires more bandwidth and can also yield undesired effects, for example, by registering irrelevant noise fluctuations in a sensor reading. Often, 128 values are all you need to accomplish a task, and sometimes even that is too much. The key is to decide exactly how many levels of resolution you need, what the high- and low-end scale of your data should be, and then use software to scale the incoming values (128 in most MIDI messages) to the right increment, minimum, and maximum.

Think about the problem musically. Here's the fun part. If you have some arrays of lights (digital videos on a computer, pyrotechnics, or whatever) and you need to control them, why not think of each as a set of musical notes? Then, the channel can determine which array of lights you want to direct (imagine each as a different musical instrument); pitch becomes the number of the lights in the array; velocity becomes brightness; and a value such as program change (which determines what kind of sound a MIDI note uses) could become color. You're conducting an orchestra of sounds, lights, motors, or anything else you can rig up to respond to MIDI messages.

MIDI is a useful and flexible tool, but that doesn't mean it's the best choice for every project. MIDI cabling imposes length restrictions, and large amounts of data can sometimes suffer from latency. There are alternatives, such as serial data or the new Open Sound Control standard (opensoundcontrol.org). But MIDI is generally the best choice when you need to make use of hardware or software that is MIDI-compatible out of the box.

Hands-on: A Sensor Software Interface

Using a sensor kit, you can turn readings from almost any sensor into MIDI data that will control a wide range of software. This is a powerful option for projects such as art installations and live audiovisual performances. Let's look at a specific example using a light sensor to control a rendered video image.

Photography by Peter Kirn

The basic setup is as follows. A photocell feeds a MIDIsense board that sends MIDI messages to Max/MSP/Jitter software running on a computer. The software generates an image that morphs between a doughnut-shaped torus and a sphere (depending on the sensor reading), which twists and contorts in the process to create an interesting visual effect (see Figures 1 and 2 on the next page).

There are many sensor-to-MIDI options, but Limor Fried's MIDIsense is a good starting point. It's an open source hardware platform that's very simple to configure and use, and it works with popular resistive sensors, including flex or bend sensors as found in the classic Nintendo Power Glove, and more common sensors such as potentiometers (known to most of us as "knobs"). For this demonstration, we'll use the light-sensitive photocell included in the MIDIsense kit; with this sensor, we won't have to worry about finding the right resistor to pair with it.

Here's what you'll need to do this:

MIDIsense board ($50 kit, ladyada.net/make/midisense)
MIDI-to-USB interface, such as the M-Audio Uno or Edirol UM-1EX ($40)
Mac OS X or Windows XP computer
Max/MSP and Jitter software (free trial version available at cycling74.com)

1. Find the 6 sensor blocks along the side of the MIDIsense board. Stick the long ends of the photocell into the screw terminals on one of the blocks. Snap a 9V battery into the board and switch it on. If you built everything correctly, the LED on the corner of the board will flash, which indicates that your board is sending MIDI data. See the MIDIsense website for details (ladyada.net/make/midisense).

2. Plug the MIDI side of the USB-to-MIDI interface into the board's MIDI in and out jacks, and connect the USB side to your computer.

3. Configure the board to send the data you'll need. The MIDIsense site has a Mac software utility, cross-platform Python scripts, and (by the time you read this) a graphical Windows-native option for this task. Using the configuration utility, turn off data for all the sensor inputs except the one into which thephotocell is connected. Select a MIDI message to assign to the sensor. You can use any CC message you like; here, let's use CC 16, defined by MIDI as a "General Purpose Slider" with a data range from 0 to 127.

4. Now the fun part. Your MIDIsense board will now send MIDI Control Change values to any software, so you could run some music software and convert this photocell into a light-sensitive synthesizer for a Theremin-like effect. But instead, let's go off the beaten path and control some visuals. Install and run Max/MSP, which is a programming environment for music and multimedia performance, and Jitter, a library of visual objects for Max/MSP. Max/MSP and Jitter let you assemble and play "patches," — virtual audiovisual instruments composed of various inputs, outputs, sources, and effects patched together.

5. In Jitter, open the patch *examples/jit-examples/render/jit.gl.gridshape-morph.pat*. This is a morphing geometric shape effect that uses the OpenGL 3D graphics libraries.

6. Check the "start rendering" checkbox and select 2 shape primitives for morphing. The "xfade amount" translates from one shape to another and, currently, gets its value from the number box above it, which you can change by dragging on the number or typing in a new number.

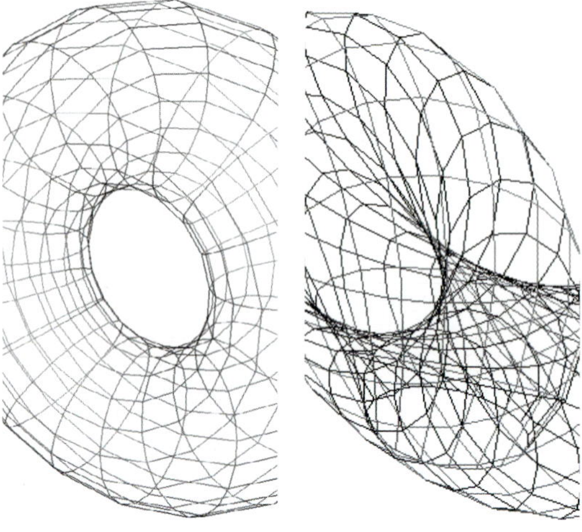

Figure 1 **Figure 2**

7. To use the light sensor to control the crossfader instead of the mouse, input the data using Jitter's ctlin object ("controller in," responding to Control Change messages). You only want data from the light sensor, so type ctlin 16 6. This specifies controller 16 on MIDI channel 6, assuming you plugged the photocell sensor into connector 6.

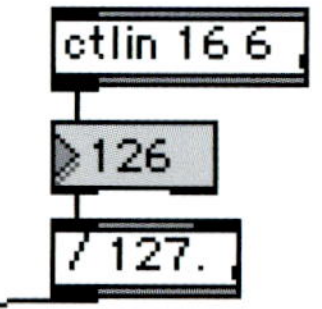

8. The ctlin object outputs data within the range of the current controller; for CC 16, that's 0 to 127. To see the results, patch the outlet of the ctlin box to the inlet on a number box object. But since the crossfade value ranges from 0 to 1, we need to scale it down. Add a new box and type in / 127. That divides any inlet value by 127, giving the range you need. The decimal point is needed after the 127 to specify a floating point result, rather than an integer. Otherwise, the number would just bounce between 0 and 1.

9. Patch the result of the calculation to the xfade amount number box, and voilà! Move around the photocell and the varying amounts of light will change the 3D morphing object. You now have sensor-controlled visuals, and you can extend this principle to any interactive art. In a future article, we'll look at a more fully developed project using sensors and MIDI.

MEET THE LEAGUE OF ELECTRONIC MUSICAL URBAN ROBOTS

1

2

3

4

LEMUR (League of Electronic Musical Urban Robots, lemurbots.org) was founded in 2000 by Eric Singer. Even though the robotic instruments are computer controlled, they play "real" (as opposed to synthesized) music, combining the warmth of real instruments with the precision of robots. (1) **ModBots**, designed by Bill Bowen and Singer, are intended to add ambient sound to architectural spaces. (2) **ForestBot**, designed by Jeff Feddersen and Milena Iossifova, is made from 25 ten-foot flexible metal stalks tipped by egg rattles. Computer-controlled vibrational motors shake the stalks. (3) Chad Redmon, Kate Chapman, Kevin Larke, and Singer's **TibetBot** uses six mallets to strike three traditional singing bowls. (4) **!rBot**, by Feddersen, Iossifova, Michelle Cherian, Brendan J. FitzGerald, Singer, Larke, and Ahmi Wolf, opens like a mollusk and rattles a bouquet of Peruvian goat-hoof maracas.

Composer and media artist Peter Kirn is the author of *Real World Digital Audio* (Peachpit Press, 2005), a book on making music with digital technology, and editor of createdigitalmusic.com.

Create ant farm documentaries, play record albums with a plastic cup, and do some pocket-sized instant messaging.

TOOLBOX

Up Close and Personal

QX5 Computer Microscope

$79 compuvisor.com/qx5diblmicom.html

The QX5 is the ultimate microscope for the casual scientist and is guaranteed to provide even the mildest-mannered geek with countless hours of fun. In addition to viewing, saving, and exporting magnified images, you can also take the microscope out of its cradle and use it as a handheld camera — extreme close-up style.

Yes, that's right, ladies and gentlemen. You, too, can now create documentaries of your ant farm, get an up-close and personal look at that nasty paper cut, or just muse to your brown-eyed girl about how pretty her irises appear when magnified 200 times.

Mounting the microscope to create time-lapse movies from still images of bacteria growing or plants sprouting is another really neat thing you can do, and the software included with the kit takes care of all the messy details of exporting your scientific research, allowing you to concentrate on the experiment at hand.

Really, my only complaint is that the software only works for Windows users, but not to worry: Linux and Mac users can seek out and/or buy additional third party software to get things rolling, and the effort is well worth it.

—Matthew Russell

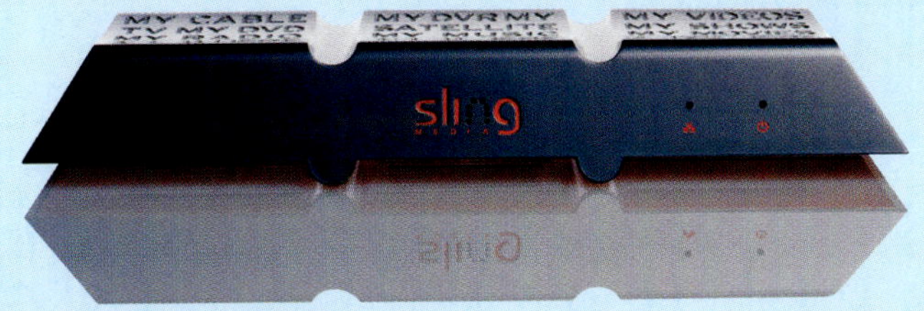

Road Warrior

Slingbox Mobile Home Theater

$199 slingbox.com

The road. Sure it can be ugly. The lumpy beds, flat pillows, and sub-par chicken wings from room service. You're desperate for comfort, even something as meager as watching your favorite TV show while waiting for room service. As long as your hotel has an internet connection and you have a Slingbox at home, you'll never have to watch The Weather Channel again.

Slingbox and the SlingPlayer, which you download to your laptop, feature support for everything from your TiVo to your cable or satellite box. Slingbox installs into your existing home setup in minutes with a cool intuitive wizard that auto-discovers and configures. You can even get a mobile version that you use on your smartphone or on Pocket PCs that use the Windows Mobile OS (a Mac-compatible version is on the way). I think I hear room service knocking.

—Rob Bullington

The Ultimate Action Figure

Acrobots

$7 thinkgeek.com/geektoys/cubegoodies/6748

Acrobots have a peculiar expressiveness that's articulate and strangely open to interpretation — and this is what makes them the ultimate action figure for the office. Use their jointed bodies and magnetic hands to erect pyramids, re-enact scenes from kung fu movies, and stage dramatic displays that hint of emotions that you just can't otherwise communicate. Believe me, there's no better way to spend a dull moment at work.

—Matthew Russell

Japanese Marking Gauge

$35 makezine.com/go/marking

▪ It comes up often enough in the shop: you want to mark a line "x" inches in from the edge of something. You end up futzing around with tape measures, pencil tick-marks, and straightedges — but still aren't sure you got it exactly right. Well, Japanese woodworkers have been worrying about this problem for a few centuries longer than you, and have evolved a better way: the marking gauges called *kebiki*.

A similar western tool exists (expensive brass and rosewood models are popular in the tool-fondler catalogs), but I admire the simple elegance of the Japanese version. The kama kebiki type has a comfortably rounded oak grip, and two (retractable) steel knives for scribing lines. Its official function is laying out mortises, but even after owning one for 20 years, I'm still discovering surprising new uses for mine. Scoring cut-lines in wood reduces chip-out, and strips of thin veneers, plastics, etc., can be sliced off directly. And it's one of those special tools that gives you pleasure just picking it up.

—Ross Orr

Editor Phillip Torrone talks about kits that caught his eye on makezine.com.

Cosmic Poppers Kit

$7 makezine.com/go/poppers

This looks hilarious (and is a simple kit to make). Once you build it, this kit randomly creates explosions of sound and light when it senses the absence of light. Halloween, anyone?

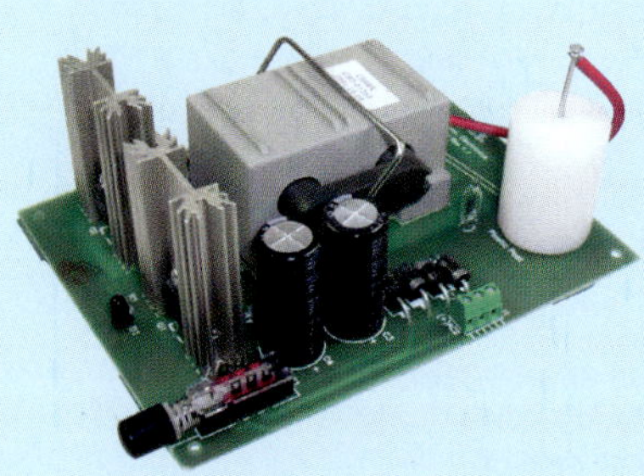

Plasma Generator Kit

$64 makezine.com/go/plasma

This looks like yet another fun kit. You can build plasma balls, light up fluorescent tubes with no wires, and make big sparks. It's a solid-state, high-voltage power supply, and perfect for high voltage experiments (even photography). And you can't even electrocute yourself, so experiment away.

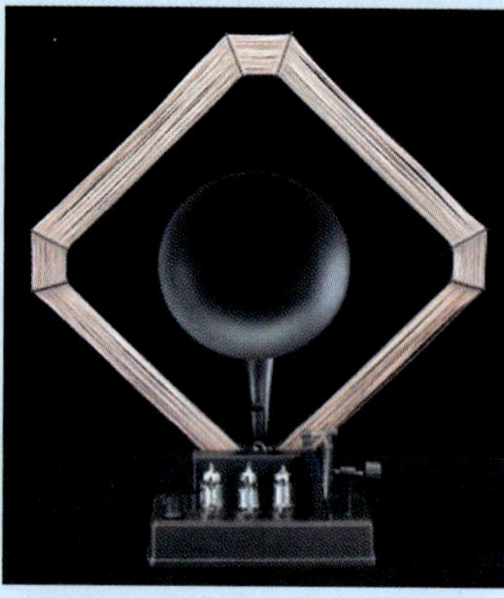

Vacuum Tube Radio Kit

$190 e-clec-tech.com/vatura.html

• You can put together your own radio with vintage vacuum tubes. Not only do you learn about basic electronics and radio, but it looks really cool.

Enigma Machine

$220 xat.nl/enigma-e/desc/index.htm

• During World War II, the Germans coded their radio messages using the Enigma ciphering machine; when the Allies finally figured out how to build one, cracking the German code, they changed the war. There are Enigma machines on eBay from time to time, but this kit looks like it would be a fun build (and a lot cheaper). It's an updated version of the classic Enigma that uses modern electronic components.

Classic Video Table Tennis Kit

$20 thinkgeek.com/geektoys/science/8546

Wow, for $20 you can make your own Pong! You just need to be able to solder, and you can turn your TV into the classic table tennis video game. (Play against another geek, or challenge the computer at four different skill levels.)

Plastic Cup Phonograph Kit

$50 makezine.com/go/phonograph

▪ **Record your voice, Edison style. The kit uses a plastic cup and a needle instead of a waxed pipe and stylus, but the effect is the same. For those of you who are into Gakken kits, VeryCoolThings has all of the Otona no Kagaku kits on its website (there are more than 20).**

Kafka Pinstriping Kit

$90 xrl.us/oerj

▪ **Great kit for pinstriping your latest computer case mod or robot companion. Designed by pinstriping master Steve Kafka, the kit shows you the basics of customizing anything you can paint.**

Henry Field's White Button Mushroom Kit

$28-$35 henryfields.com

▪ **Imagine a room full of 3- and 4-year-olds pulling on you, tugging on you, yanking on you — all desperate to show you something. That's what my life was like when my son's preschool got this nifty kit. For about 30 days I heard all about it — the daily misting with the spray bottle ("Mom, Vega got to do it today."); watching the stringy mycelium spring into being ("Now it's growing spider webs!"); and finally, the mushrooms themselves ("Some are big, some are small. I don't like to eat mushrooms, though.")**

—Shawn Connally

Tube Time

Nixie Desk Clock Kits

$99-$250 tubeclock.com

Nixie tubes are vacuum tubes that contain filaments representing the digits 0-9. These beautiful relics of early computing display their numerals with a quivery orange glow. Today, you can buy a Nixie tube clock kit with parts and instructions from Peter J. Jensen of Ann Arbor, Mich.

It was a joy to put the clock together, thanks to the clear step-by-step instructions and parts envelopes labeled with numbers corresponding to the numbered instructions. Jensen designed the clock and circuit board himself, and his elegant sense of aesthetics became apparent to me as I soldered the components onto the board and mounted it and the tubes into the handsome metal case. The finished clock looks like the creation of some famous mid-century modern designer. I plan to buy more of Jensen's clock kits to make and give as gifts.

—Mark Frauenfelder

FAVORITE THINGS

Chosen by the Institute for the Future

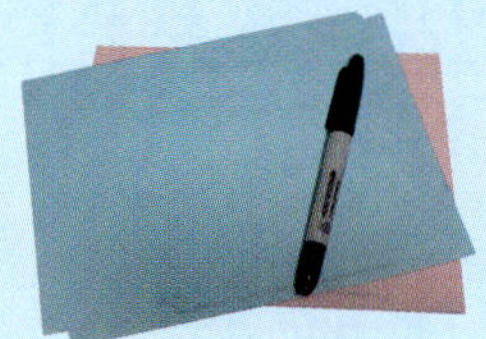

Jumbo Sticky Pads: These giant sticky pads (about 6"×8") are perhaps the secret sauce of the Institute. In group brainstorms, people write one short and simple idea, thought, or question onto a single sticky, and the generous size means there's plenty of space to WRITE BIG. Since the cards are re-positionable, it's easy to mix and mash to form new threads. The same technique works solo, too.

***Found* Magazine:** To better communicate our research, we've started creating all sorts of provocative products and ephemera from the future, and there's no better source for the minutiae of daily life than *Found*. Each issue is a voyeur's paradise, chock-full of lost letters, scribbles, to-do lists, dropped bills, photos, you name it — and most of the submissions are sent in by actual readers. Browsing this random collection is a great way to unplug and remember how raw and messy human existence really is.

TAP Plastics: As we learned from *The Graduate*, "There's a great future in plastics," and odds are that if it's made out of acrylic, TAP Plastics sells it, from tubes and hinges to boxes and sheets, in almost any color (and opacity) of the rainbow. Buy the raw materials and DIY, or they'll build it for you. Check out their custom formulations of epoxy and resin too, made just for prototyping. The website is also a great source of project ideas. Check out tapplastics.com.

Xyron Adhesive Machines: When you need to get a relatively flat thing to stick to something else, and spray mounting is out of the question, this is your best bet. Just slap down the item, turn the crank, and out it comes with an even application of gunk on one side, much dryer and sturdier than spray mount. Xyron machines come in multiple sizes: the bigger machines are expensive, but many printing and copy shops have one and charge by the foot.

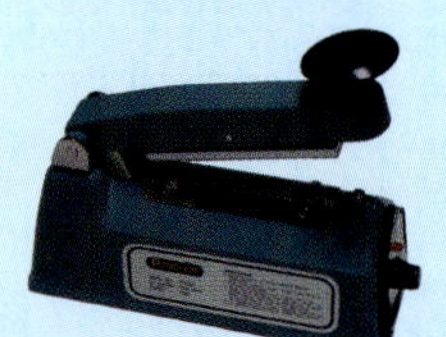

Heat Sealers: This gizmo really appeals to your inner Beavis' desire to burn and mutilate things. We bought ours to mock up a fictional product, but come up with new uses for it all the time. Basically, put two pieces of meltable material under the arm, press down, and voilà — a new sealed seam. Like Xyron machines, these get pricey for large models, so try befriending an employee of a packaging or shipping store.

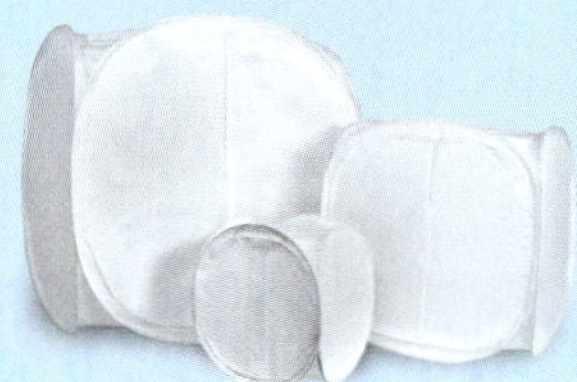

Photo Shooting Tents: Once your masterpiece is ready to take the future by storm, you'll need great photos of it. Shooting tents will cradle your small-to-medium-sized creation on a solid background and evenly diffuse the light coming in, particularly important for objects with reflective surfaces. There are a lot of kits that include lights and stands, but if you already have decent lights or a good source of natural light, then just buy the tent itself.

The Institute for the Future (iftf.org) is a nonprofit strategic research group with over 35 years of forecasting experience. Their core work is identifying emerging trends and discontinuities that will transform global society and the global marketplace. They provide their members with insights into business strategy, design process, innovation, and social dilemmas.

Hear the Action as It Happens

Uniden BCD396T

$799 (but often discounted) uniden.com

Remember those old crystal-controlled police scanners it seems just about everyone had during the CB radio craze of a few decades ago? You probably have visions of an old tabletop scanner, with its flashing bank of glowing red LEDs and the crackle of your local fire or police departments being heard through its speaker, sitting next to your dad's favorite recliner. Well, communications technology has rapidly progressed over the years, and so too has radio scanning equipment. With advances that seem virtually light years ahead of the scanners of days gone by, the Uniden BCD396T offers capabilities unheard of just a few years ago.

With coverage of just about every frequency from 25MHz to 1.3GHz (sorry, the cellular bands are excluded as per the FCC), the handheld BCD396T lets you tune into police, fire, weather, ambulance, aircraft, military, amateur radio, marine, news media, and even FM and VHF-TV commercial broadcasts. Sporting a memory capacity of up to 6,000 alpha-numeric channels, the PC-programmable and controllable BCD396T utilizes Uniden's "TrunkTracker IV" technology, allowing you to follow the action of most trunked radio systems commonly used by emergency response agencies. More than 400 major U.S. cities are preprogrammed into the radio for rapid deployment.

In addition, the BCD396T also features Uniden's "Close Call" RF capture technology, which allows the scanner to lock onto nearby transmissions even if you haven't programmed anything into your radio. Other features include fire tone-out capability, Weather Alert, Search and Store operations, and Priority Scan.

This definitely isn't your father's scanner. *—Joseph Pasquini*

Let There Be Light

Screw-base Fluorescent Lights

$5 makezine.com/go/bulb **or** en.wikipedia.org/wiki/Compact_fluorescent

If you spend your workday in Cube Hell, the very words "fluorescent light" might send you off screaming. But there's been a quiet revolution in compact fluorescents lately: no deathly blue pallor, no hum, no flicker. The light is warm and slightly pinkish; the form factor is a drop-in replacement for old-tech incandescent bulbs. But who would buy a $5 light bulb?

You will, once you do the math: you can save more than $50 over the lamp's lifetime in electricity. And since they last ten times as long as incandescents, you'll save a bit more on bulb replacement too. But they're best for locations where they'll stay on for a few hours — frequent on/offs can burn them out earlier than their rated lifetimes.

—Ross Orr

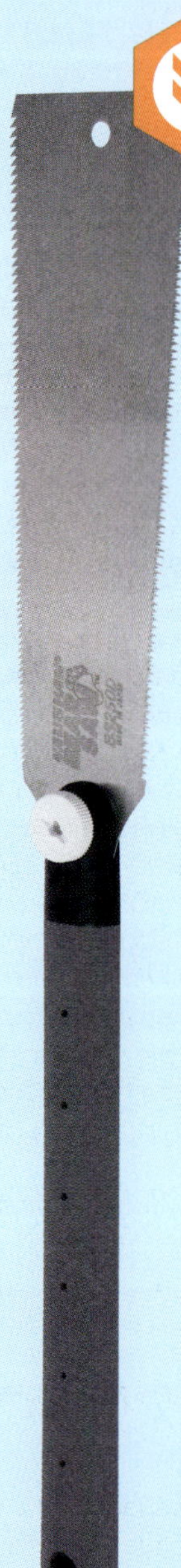

Turning Japanese

Vaughan BearSaw

About $20, available at most hardware stores

vaughanmfg.com/bearsaw_frameset.html

When I think of Japanese tools, I think of minimalism and efficiency, two qualities perfectly embodied by Vaughan's BearSaw, actually modeled after a traditional Japanese saw called a *ryoba*. For years I used a regular old crosscut handsaw, which worked fine — unless I wanted to cut a piece of wood with the grain. Or I needed to cut something flush. Or I needed to make a delicate cut on a fragile piece of wood. It got to the point where cutting something was a fearsome process.

Then I found the BearSaw. It has a crosscut side and a ripcut side; it cuts on the pull stroke, so the blade is thinner, more flexible, easier to control, and doesn't remove as much wood. It also has gradated teeth (there are more teeth per inch closer to the handle) for making delicate cuts. Best of all, the blade is removable and replaceable, which means I can fit a full-size saw into my portable toolbox. I still have a use for my old crosscut saw, though. I got a bow and I'm learning to play it as a musical instrument.

—Ty Nowotny

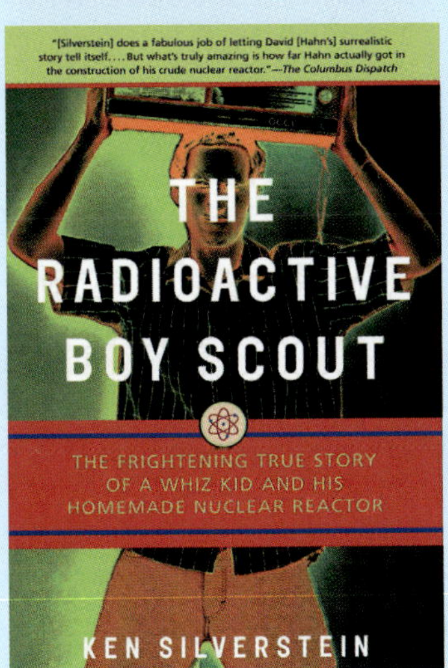

Golden Boy

***The Golden Book of Chemistry Experiments* by Robert Brent**
$100-$700, Golden Books (out of print)
***The Radioactive Boy Scout* by Ken Silverstein $14, Villard Press**

A few years ago, I read a bizarre and wonderful article in *Harper's Magazine* by Ken Silverstein, called "The Radioactive Boy Scout." It was the true story of David Hahn, a boy who ended up collecting enough radioactive material that a Geiger counter could detect it from five houses away. In January 2005, Silverstein published a book based on the article, called *The Radioactive Boy Scout: The Frightening True Story of a Whiz Kid and His Homemade Nuclear Reactor.*

The article and book mention that young David had been inspired to practice chemistry after receiving *The Golden Book of Chemistry Experiments*, which includes experiments on making chlorine, ammonia, hydrogen, and ethanol. The book is long out of print, and used copies are very expensive. In today's litigious environment, no major publisher would dare republish a book that had actual chemistry experiments in it.

I have long wanted to own a copy of *The Golden Book*. A friend emailed me a scanned page recently, and it prompted me to search for a sub-$100 copy. I got lucky and found a free copy of the 112-page book, thanks to BitTorrent (mininova.org/tor/116968).

The book is an example of everything great about vintage children's science books. Once you lay your eyes on it, you will come to the sad realization that our society has slipped backward in at least three important ways: 1) The writing quality in old kids' science books was better, 2) The design and illustrations were more thoughtful and skillful, and 3) Children in the old days were allowed and encouraged to experiment with mildly risky but extremely rewarding activities. Today's children, on the other hand, are mollycoddled to the point of turning them into unhappy ignoramuses.

I can't wait to conduct these experiments with my daughters at my side.

—Mark Frauenfelder

Damn Good Design

***Design Like You Give a Damn*, edited by Architecture for Humanity**
$35, Metropolis Books

Design Like You Give a Damn is a refreshing variant on the architectural coffee table book, a fabulous and stimulating treat for the social-minded maker. True to Kate Stohr and Cameron Sinclair's philosophy of the open sharing of ideas, it's an artfully laid out book presenting architectural and design solutions for disadvantaged communities. They start out contextualizing the social design movement, and then move on to examples. From temporary housing for refugees made out of paper logs, to plastic-bag shelters that connect to external heating ducts for the homeless, to a water pump powered by children at play, these are out-of-the-square solutions for fabricating new worlds.

—Saul Griffith

Just Zipit, Zipit Good

K-Byte's Zipit Wireless Messenger

$99 zipitwireless.com

K-Byte's Zipit Wireless Messenger is a cute pocket-sized Instant Messaging (IM) appliance that can connect to any wi-fi access point for on-the-go IM. Just turn the Zipit on, sniff out a wireless network access point, and start messaging through your usual IM account (AOL, Yahoo, Microsoft, etc.). The Zipit even automatically scans for available wi-fi access points.

That would be the end of the Zipit story if it were not for one unbelievable feature hidden on the device's main circuit board. Along with the 320x240 monochrome LCD, and 16MB SDRAM, is a 2MB flash memory chip dedicated to storing the Zipit's Linux ARM 2.4.21 kernel. Sing hallelujah! That's a usable Linux OS inside a half-pint portable wi-fi computer that costs less than $100.

Of course, to savor the full exciting potential of this Linux installation, you must hack the little Zipit. (Before starting, make sure that you fully understand and accept the possibility that you can "brick" the device or render it useless with just one boo-boo during the wi-fi reflashing procedure.) See makezine.com/07/toolbox for the best Zipit Wireless Messenger hacks, starting with the basics. Zipit programming is in its infancy, so more applications can be expected, and contributions are welcome.

—Dave Prochnow

+ See more resources at makezine.com/07/toolbox.

Progressive Pan Scraper

$2 for two, organize.com/panscrapbypr.html

Breathe easy. The problem of the ages (by which I mean getting those congealed eggs off the bottom of your pan after you've forgotten them in the sink all week) has been solved. My latest crush is this simple plastic pan scraper, perfectly shaped for getting into all sorts of hard-to-reach curves. Not only does it remove the hardened remains of last night's dinner without damaging the pan (even Teflon is safe), I've found it handy for a myriad of other uses, including scraping ice off a car windshield when the regular scraper went AWOL, and removing old paint. Now I just need to attach it to my keychain, and it can come with me everywhere I go.

—Arwen O'Reilly

Orange You Glad?

Citra-Solv Natural Solvent

$6 citra-solv.com

Worried about the toxicity of turpentine? I just discovered that highly distilled citrus oil (available under such brand names as Citra-Solv and Natural Solvent Spotter) takes off fresh oil paint just as well, leaving behind a strong scent of orange or grapefruit peel. It's also great for taking off sticker residue and even chewing gum. It can be used to take grease and oil spots out of clothes, but I've discovered that it leaves a ring on some fabrics, so for that job, I prefer using talcum powder. (Just pour talc on the spot, leave overnight, and brush off in the morning. It will lift the grease out like magic, as long as you actually brush the powder out thoroughly with a stiff brush.)

—Tim O'Reilly

Slice of Life

Apple Peeler-Corer-Slicer

$25 amazon.com/gp/product/B0000DE2SS/

Peeling and cutting up the apples was always the barrier to entry for me when I considered making an apple pie (especially since one is never enough) — until I discovered a curious device called an apple peeler-corer-slicer. A Wallace and Gromit-style device if ever you saw one, but it works like a charm.

Just stick a firm apple onto the triple spikes, turn the hand crank a half dozen times, and the apple is cored, peeled, and neatly sliced in a 1-inch-wide spiral, with a longer spiral of peel spilling over the side of the counter (hopefully into a bucket or bowl artfully placed below, but if not, onto the floor). Make a single downward cut with a knife, and you have a set of neat rounds perfect for pie.

Those neat rounds inspired another great apple pie innovation in our house. The holes in the middle were so tempting that I decided one day to drop in raspberries. A perfect variation on the traditional apple pie.

Another tip: Either use sweet apples like Fujis and use no sugar, or use sour apples like Granny Smiths and sugar lightly. Most American apple pies are way too sweet. My ideal apple pie is tart.

—Tim O'Reilly

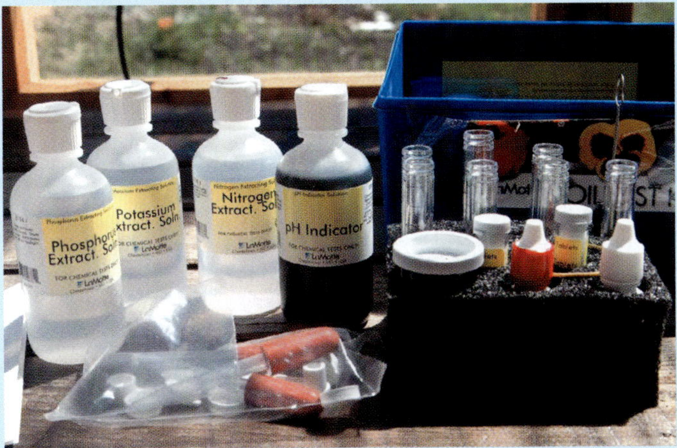

Pay Dirt

LaMotte Soil Testing Kit

$60 territorial-seed.com

Wonder why your tomatoes aren't taking off? Soil Test kits allow you to learn more about how your garden grows. This particular kit lets you test soil samples for concentrations of nitrogen, phosphorus, potassium, and soil pH. Once you have the test results, you will know how to amend the soil, which can improve growing conditions for your plants.

LaMotte kits are widely available in slightly different configurations, but I got mine online from Territorial Seed Company. The kit includes several informative booklets on soil science.

—Dale Dougherty

Matthew Russell tries hard to live life as a renaissance man.

Dave Prochnow writes about his projects at pco2go.com.

Joseph Pasquini is an avid amateur radio operator as well as a shortwave and scanner listener.

Ross Orr keeps the analog alive in Ann Arbor, Mich.

Tim O'Reilly is founder and CEO of O'Reilly Media, Inc.

Ty Nowotny is a full third of the triumvirate of havoc-wreaking engineering interns at the MAKE Lab.

Saul Griffith works with the power nerds at Squid Labs.

Rob Bullington is the marketing and events coordinator for MAKE.

Have you used something worth keeping in your toolbox? Let us know at toolbox@makezine.com.

RETROCOMPUTING

Digital Spelunking

By Tom Owad

Unearthing ancient Apple II files on AOL.

The ephemeral nature of the web has appalling implications for the preservation of information. Even many once-common Mac OS 9 shareware apps are now impossible to find, to say nothing of older and more obscure programs. How many web and FTP sites from 1994 still exist?

In the days when BBSes were more popular than the internet, America Online's file libraries rivaled the entire internet in their breadth of files. It was in an effort to determine whether these ancient libraries still existed that I gingerly installed AOL on a spare computer and signed up for an account.

My hope was that even though the original keywords for these libraries no longer worked, I might still be able to find a link or URL that would take me to the old file library search function, much in the way that it's still possible to access the Apple II forum on AOL by going to keyword: aol://4344:1264.a2main.10029531.514525857.

After much effort, I failed to find the search page, but I was able to use the Apple II libraries to understand the URL syntax. The URL for the Apple II New Files library is aol://4400:8287, and here is the URL for the UnForkIt file, contained in that library: aol://4401:8287:636250. The first value identifies the resource type; in this case, either 4400 for a library, or 4401 for a file. The second number, 8287, is the library ID, and 636250 is the file ID. The file IDs are not consecutive within libraries.

By changing the library ID, it is possible to access file libraries that no longer have direct links. I tried random IDs and was soon bookmarking the Connectix Macintosh Library and MacHacks.

As AOL has moved away from its proprietary Rainman language in favor of HTML, and sophisticated users have left for the internet, these old areas of AOL have been abandoned and forgotten.

Each file records the date it was most recently downloaded. In many of these libraries, the most recent download was from 1999. In a few, the records have been confused with those for other files — the description will be for a shareware utility, but the file will be a JPEG. Other file downloads fail halfway through. The system, it appears, has been left to decay.

But the bulk of the files I found were in salvageable condition. The challenge was to find the surviving software libraries and download their contents. I wrote a QuicKeys script to open the Keyword window, type in a possible library address, and bookmark the address if it worked. Over a period of several days, this script tried about 70,000 possible library locations. Another QuicKeys script is now visiting each library and downloading the files. If you'd like to learn more about the project, or lend a hand, visit applefritter.com/aol.

Tom Owad (owad@applefritter.com) is a Macintosh consultant in York, Pa., and the editor of applefritter.com. He is the author of *Apple I Replica Creation* (Syngress, 2005).

Illustration by Kirk von Rohr

Puzzle This

By Michael H. Pryor

MAKE's favorite puzzles. (When you're ready to check your answers, visit makezine.com/07/aha.)

Jelly Beans

There are three jars in front of you that are all mislabeled. You can't see inside the jars, but one contains red jelly beans, another holds blue jelly beans, and the third has a mix of both (not necessarily a 50/50 mix; it could be a 1/99 mix, a 299/22 mix, etc.). What is the least number of jelly beans you would have to remove from the jars to figure out how to fix the labels?

Mountain Man

At 6 a.m., a man starts hiking up a mountain. He walks at a variable pace, resting occasionally, but never actually reversing his direction. At 6 p.m., he reaches the top. He camps out overnight, and the next morning at 6 a.m., he begins his descent down the mountain. Again, he walks down the path at a variable pace, resting occasionally but always going downhill. At 6 p.m., he reaches the bottom. What is the probability that during the second day, the man is at a particular spot at the exact same time he was there on the first day?

River Crossing

Three cannibals and three anthropologists have to cross a river. The boat they have is big enough for only two people at a time. If at any point there are more cannibals on one side of the river than anthropologists, the cannibals will eat the anthropologists. The cannibals will follow all directions given to them by the anthropologists. So what can the anthropologists do to ensure they do not get eaten? Remember, the boat cannot cross the river by itself, so someone has to be in it to row it across each time. And as soon as the boat arrives on the shore, the number of cannibals must be less than or equal to the number of anthropologists on that side.

Illustrations by Roy Doty

Michael Pryor is the co-founder and president of Fog Creek Software. He runs a technical interview site at techinterview.org.

BLAST FROM THE PAST

Workshops, Big and Very Small

By Mister Jalopy

Build a stowable mini workshop for modest tasks like lamp rewiring and scissor sharpening and soon you'll be building your dreams.

As sensible people slumber, the palpable thrill of junk acquisition wakes me at 6:30 a.m. every weekend. Sleeping on Saturday mornings is for those who have not fallen prey to the temptress of garage sales. There are such unbelievable riches buried in suburban cul-de-sacs that I sometimes wonder if I'll find the Hope Diamond in a junk drawer amongst the S&H Green Stamps, unsticky tape, and keys without locks.

When arriving at an estate sale, I forgo the house and go directly to the under-lit, spider-webbed garage. Time is frozen under a layer of dust, and you can get a feel for the makers who preceded you with their tube radios, home weather stations, and oddball tricks.

At their best, garage workshops are chock-full of raw materials like baby food jars filled with cup hooks, cigar boxes of lamp parts, reclaimed kitchen cabinets full of solidified cans of Plastic Wood, and neatly coiled power cords cut from broken appliances. In the postwar, cardigan-wearing and pipe-smoking handyman heyday, considerable creativity, planning, and reuse trumped money when working on the project that would never end — building the ideal workshop.

Based on my extensive field research, it seems home workshops assembled today are but a whisper of what they used to be. Of course, the workshops have changed along with the world. The external forces of Depression-era poverty and the following wartime rationing shaped a generation's view of what was worth fixing, saving, and reusing.

At some point, the cost of buying a new hammer dropped below the cost of replacing a broken hammer handle and everybody was able to afford a hammer. Then came the economical pocket calculator, and now a $200 hand-crank-powered computer is right around the corner.

Affordability is a great technological advance in itself and empowers those who previously could not afford a hammer, calculator, or computer, but there are losses when replacing instead of fixing. Besides the chain of raw materials needed, energy required, distribution transport, and end disposal, life is just less rich when everything you own is only five years old.

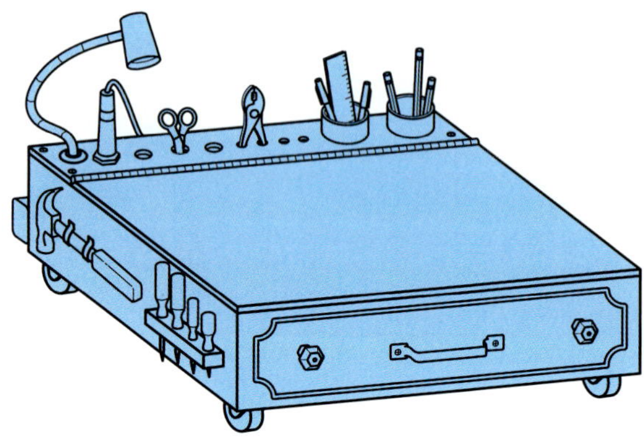

Make It Personal

Can it be improved upon? Think how pinstriping would dress it up! Or a deep-sea diver fighting a giant lobster painted on the side. Or the correct answers from last week's crossword written on the inside lid. Make it as personal as your diary.

My shop grew from a single Craftsman workbench that I couldn't access until I pulled the car out of the garage. A workbench can be as unattainable a luxury to a citizen of Manhattan as a computer is to a resident of Zambia. Inspired by the handyman magazines of the past, Mister Jalopy's Hide-Away Workbench is a modest workshop that can be tucked away until project time.

Start with your workshop and then build the rest of the world around you.

References

"Closet Door Workshop," *Science and Mechanics*, December 1952

"Better Ways to Build Workbenches," *Amateur Craftsman's Cyclopedia of Things to Make*, 1937

Mister Jalopy breaks the unbroken, repairs the irreparable, and explores the mechanical world at Hooptyrides.com.

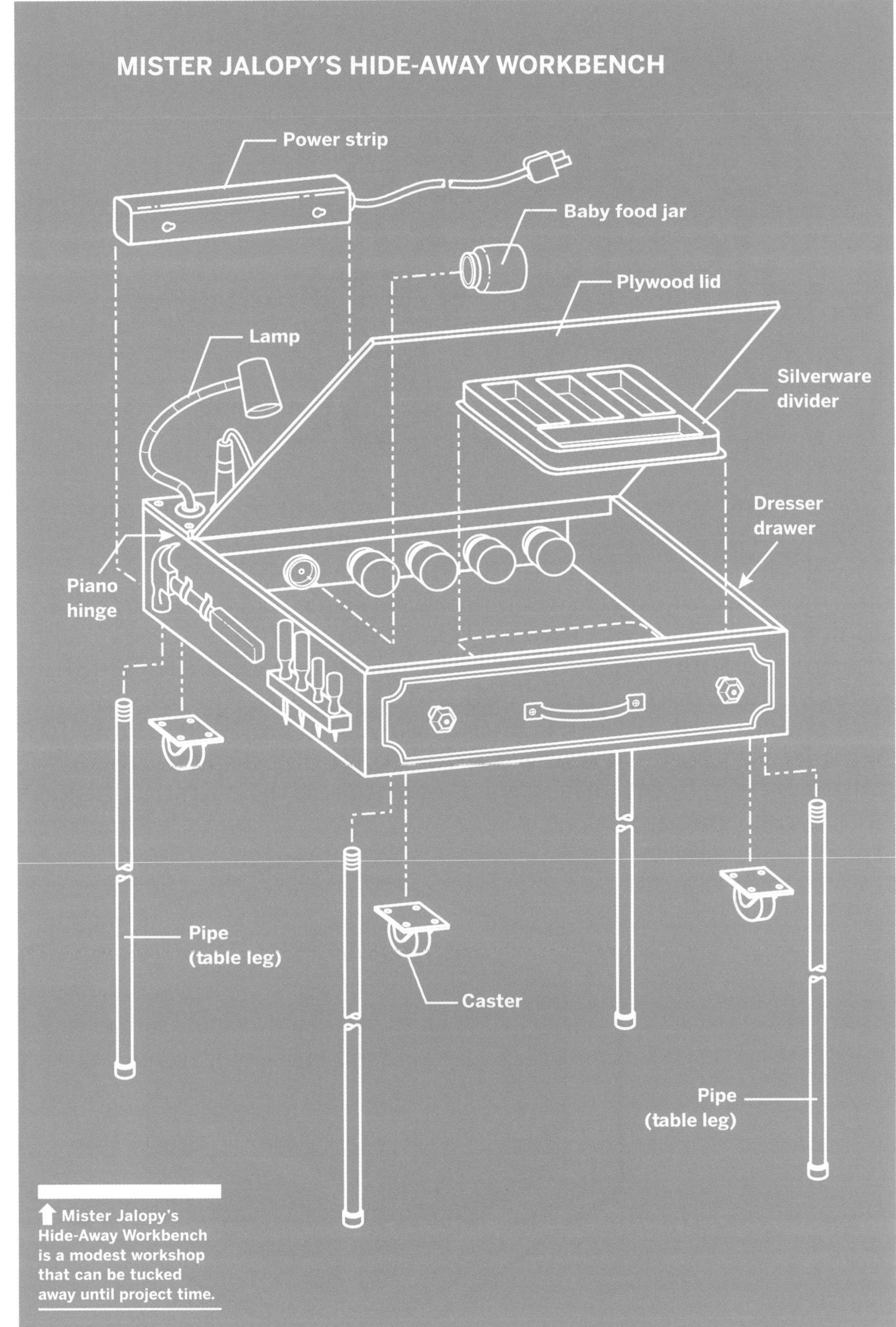

Mister Jalopy's Hide-Away Workbench is a modest workshop that can be tucked away until project time.

Illustrations by Tom Giesler / tomgiesler.com

Kid Safety Labels We Want to See

Gever Tulley, who organized a tinkering workshop for kids at Maker Faire (you can watch videos of the workshop and download instructions for his bottleboat submarine at tinkeringschool.com/blog), has been talking to me about a book idea of his that I love: *50 Dangerous Things You Should Encourage Your Child to Do*. He and I agree that American kids are raised in an overly cautious manner, out of fear that they might get hurt, and we are limiting their ability to explore a wider range of experience. In place of traditional safety warnings on kids' toys, Gever came up with the labels on the next page.

—Dale Dougherty

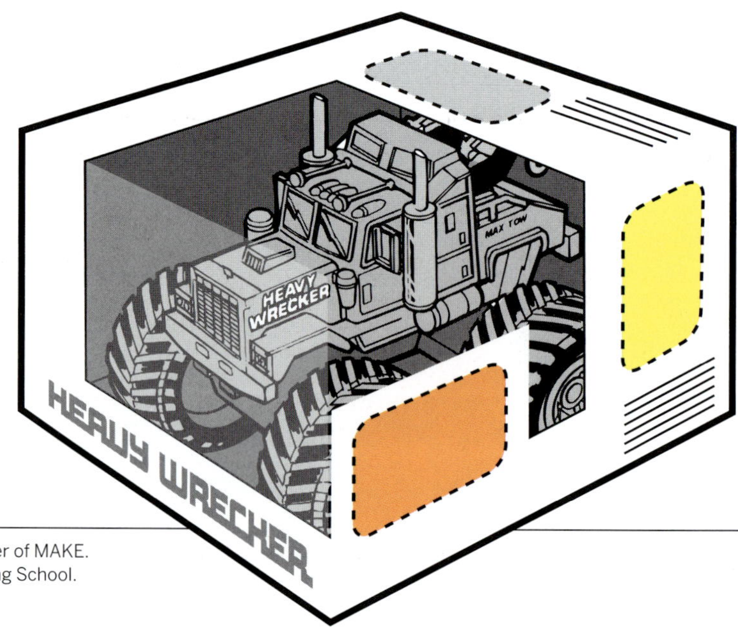

Dale Dougherty is editor and publisher of MAKE.
Gever Tulley is the founder of Tinkering School.

EARLY BRAND EXPOSURE

The maker of this toy is trying to create a positive branding experience. This brand exposure will create a lasting impression that will affect your child's behavior over the next 18 years and beyond. By exposing kids to brands rather than experiences, you will help mold their thinking and hone their consumer receptiveness. Studies have shown that these toys, with their bright colors and easy-to-use features, produce uniformly under-performing children who later become credit card abusers.

CAUTION

REDUCTION OF EXPANDED PLAY OPPORTUNITIES

The manufacturer of the toy will urge you to quickly eliminate the packing materials that this toy is shipped in. They will claim that this is for the sake of your child's safety, when in reality what they are concerned with is the very real possibility that the packaging is more interesting than the toy. Should this be the case, we encourage you to let your child have the box, the styrofoam, and the plastic bag marked "Danger — Suffocation Hazard." Rest assured that under your benevolent eye, the only thing in danger of suffocation is your child's imagination when you leave him with only the boring contents to play with.

COMPARTMENTALIZATION OF EXPERIENCE

Commercially produced plastic toys create a narrowed interaction expectation on the part of the child. The impossibly clean lines and limited play modes offered by these mass-produced, reduced-risk-of-injury designs establish boundaries that your child will quickly adapt to and accept. This mode of limited "play" will affect how your child approaches problem solving and undermine her natural curiosity.

READER INPUT

Where makers tell their tales and offer praise, brickbats, and swell ideas.

I just inherited a friend's old Sony Clie (PEG-SJ20/U) and it got me thinking: this thing is basically useless now. With all the phone/PDAs, iPods, and iPod-esque devices out there playing songs, video, and taking pictures, what is to become of all the countless older gadgets (like mine) that serve strictly as personal organizers? I'm wondering if it might make a good story to offer suggestions for hacking these older machines into something that's still useful. Hell, the thing's got a memory stick expansion slot. That's got to be good for something. It even has an IR sensor; maybe it could become my next TV remote? I'm praying to the über-nerds to resurrect this relatively inert technological artifact. Any suggestions?

—Russell Meyer

Post your suggestions at makezine.com/07/input.

My recent MAKE magazine showed up at my house yesterday, and I love the fact that I can share information provided in your magazine with people, but the magazine I received was more than slightly damaged and obviously thumbed through. The plastic was open and the pages bent. The first five magazines I received were flawless, but now Volume 06 is damaged.

I know this is an odd request, but can MAKE be shipped like it is porn? The plain brown packaging? The nondescript label?

It might save me from having the next issues of MAKE not make it to me in good shape. And then me having to write a letter and request it be shipped like it is nondescript porn.

—Chris Turvey

I cannot possibly rant enough about how fine your magazine is, so I'm a bit reluctant to even begin, for fear of not doing it justice. To help me out, please imagine at least an hour of effusive gushing.

Now then ... I must add that it's simply priceless to find out one is not the sole member of some bizarre, mutant species of über-geek/tinkerer/amateur mad scientist. MAKE readers are my people (sniff)!

And if I could somehow retire, live another 100 years, and have enough money to dedicate to the

Thank you very much. I now understand addiction. Just like a crack addict, I get a hint there is something around. Frauenfelder posts on BoingBoing. My adrenaline surges a bit. The new issues are sent!

Then, like a kid waiting for the Sea-Monkey packets to arrive, I check my mailbox every day. Is it here? Is it here?

Lies, I tell myself lies. When it comes, I will ration it out. I will only read an article or two every day. I'll *try* to make it last as long as I can. Total lies.

Is it here yet? Denial. I look in the mailbox, but no, not here yet. I pretend I knew it would not arrive. I deny my excitement.

Just a little — you won't get addicted. Good stuff, pure!

Then one day, today for example, I get home from work, and there it is. In the mailbox. Wrapped up and shiny. Virginal. Untouched. *Yet*.

I will ration it out. No, seriously, this time, I can do it.

The plastic bag comes off and goes into the recycling bin. I smell the fresh paper. I examine the cover. I close off all my work. I make sure "She-who-must-be-obeyed" doesn't have anything for me to do. Really, one article, only one.

I go out to the garage. Move the motorcycles around, clear the model airplanes off the bench, clean up the electronic speedometer I am redesigning from minivan to motorbike, oh, and put away the MIG welder that has been out too long. While I am at it, clean up the oxy-acet rig.

Last. I have to make it last. Maintain boy, maintain.

Brush down the bench. Turn on that fluorescent overhead, sit on the shop stool, and it is all over. Four or five hours later, my mind is reeling. I have a million different ideas, six more top-of-the-list projects, two people to email, a trip to OSH is coming.

I've polished off two liters of lemonade and a quarter pound of chips. I have read every page. Some pages twice. My eyes are bloodshot, and my butt is sore. Hey, that is not a comfy shop stool.

But I am reveling in the wondrousness of MAKE. There is something about paper. The internet just isn't the same. I have the boxed set. I'm working on the second year. I love your magazine. Thank you, thank you, thank you.

But Hellfire and Dalmatians, I surely feel like an addict.

—Charles Statman

MAKE lifestyle (deep sigh), I would finally consider my life meaningful.

So, quite seriously, here's to the often-endangered human spirit of resourcefulness and ingenuity which MAKE helps to nurture in a critically important way. Long live MAKE — and please don't let anyone mess with it!

—Dr. Les Garwood

I absolutely love this magazine! After reading just two issues, I find myself looking at everyday objects in whole new way ... namely: "How can I make that {blank} do something it was never intended to do?" You're an inspiration!

In Volume 06, I was pleased to see the article on kite flying, which has been a hobby of mine for over ten years. The article on tensegrity was also interesting in light of the fact that Marc Rickets and his Guildworks Studio (guildworks.com) has been using tensegrity concepts to build beautiful and amazing kites for many years. For me, kites have always represented the near-perfect fusion of art and science, and Guildworks is the epitome of this idea.

Keep up the great work on MAKE. I look forward to reading it (and maybe tackling a project or two) for many years to come.

—Doug Janelle

Just got my copy of Volume 06 and was blown away by the wonderfully detailed workmanship of Tyler Rourke's recreation of the Einstein Amplifier.

Aside from the extremely detailed chassis work, the rest of the amp does not seem to use many exotic or hard-to-find parts (I can readily identify a 5U4 rectifier, two Sovtek 2A3 power tubes, etc.). I am an engineer and tube tech, and this amp could be easily built on a breadboard in a week or so of evening work.

Could MAKE ask Tyler to share the schematics with the rest of the maker community? This would be a wonderful way to keep alive both Einstein's memory and his appreciation of music (he was a skilled amateur violin player) as well as that of the engineer who designed it for him.

—José Korneluk

Editor's Note: Several readers wrote in asking for the schematic for this amplifier. Generously, George Dyson and Tyler Rourke allowed us to publish it. You'll find it online at makezine.com/06/homebrew.

First, absolutely love the magazine. I was in such a hurry for my first issue that I sent an email asking about delivery and your kind folks sent me an extra issue almost overnight. Never mind the great support, the spare issue almost started a fight among my fellow inventors when it was put up for first-come-first get! Anyway, here's my first stab at an electric-powered vehicle (see picture).

Basics: Plastic Adirondack chair from Target, with headlights, keyed ignition, speed control in the pistol grips, foot pedal steering, under-armrest neon lighting, etc. Cruises the neighborhood at about 12 mph. Surprisingly comfortable with aircraft-like (rudder) steering footpegs.

Thanks for doing a great service for the inventors of this planet.

—Jeff Large

MAKE AMENDS

In Volume 06, the LED Throwies article on page 116 lists the wrong size magnets. The size should be ½" diameter by ⅛" (not 1") thickness.

In Volume 06, page 129, the article on how to add a front light to a Game Boy Advance makes mention of LED part #G155b2 from Goldmine Electronics, which is incorrect. The correct part number is #G15562.

In the Tensegrity project in Volume 06, page 100, the Materials list suggests using 5m of stretch cord. A number of readers wrote in to recommend buying at least 9m, or two spools of stretch cord.

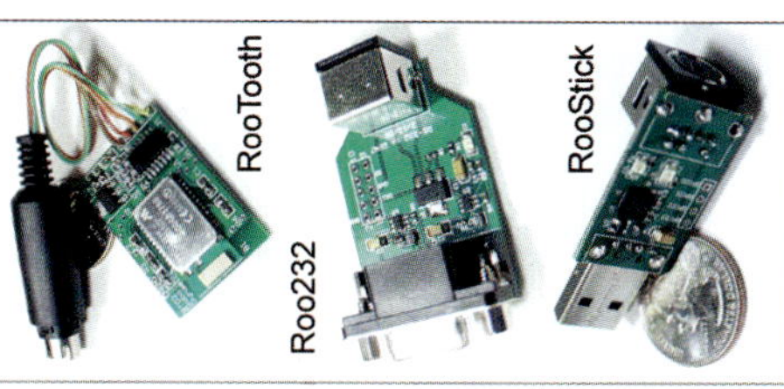

Introducing
RoombaDevTools.com
by RoboDynamics

- Development tools for the Roomba platform
- Bluetooth, USB, and RS232 using Roomba Open Interface
- General purpose robotics control boards
- Information, message boards, and tutorials.

Enabling you to create powerful robots.

ROBO DYNAMICS

HOBBY ENGINEERING

Robotics
Electronics
Science
Tech Hobbies
Metal Working
More ...

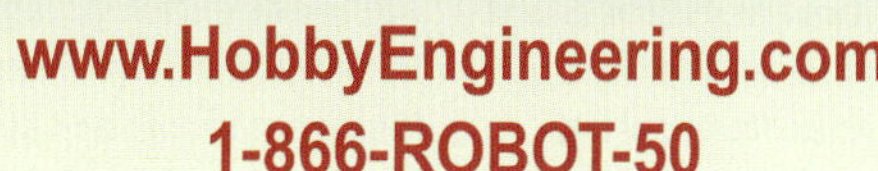

www.HobbyEngineering.com
1-866-ROBOT-50
1-650-552-9925
180 El Camino Real, Millbrae, CA 94030
Visit our store near SFO!

Project Possibilities

Some applications of the MAKE Controller that you may be seeing in the near future.

Drink-O-Mat
Timed control of valves and motorized swizzle stick can dispense and mix measured amounts of different beverages from inverted bottles.

Plant Cyborg
Light sensor and motor controllers position houseplant for optimal growth, while moisture sensor triggers valve to ensure perfect watering.

Glue Gun 3D Imaging
3-axis stepper motors and trigger controller render stable 3D forms by building them up layer-by-layer using hot glue.

Automated Pet Feeder
Motor device feeds pet remotely via web-based interface or timer program, and dispenses treats when pet stands in a specific position.

Liquid Temperature Controller
Thermostat feedback loop keeps liquid within a specified temperature range — for brewing, aquariums, candy-making, photo developing, etc.

Global Sensor Network
Large-scale survey system registers sensor input from hundreds of physical nodes connected to a common server, which publishes sensor info.

The MAKE Controller Kit

makezine.com/controller/

MAKER'S CALENDAR Compiled by William Gurstelle

Our favorite events from around the world — September – October – November 2006

September

» Burning Man
Aug. 29–Sept. 4, Gerlach, Nev. Countercultural festival in Nevada's Black Rock Desert with interesting technology, pyro-arts, machine art, and a lot more. burningman.com

EDITOR'S CHOICE
» Crazy Horse Monument Night Blast
Sept. 6, Crazy Horse, S.D. A spectacular nighttime explosion lights up the mountain with dozens of fireballs and specially designed pyrotechnical features. crazyhorse.org/events/nightblasts.shtml

» Oceana Airshow
Sept. 8–10, Virginia Beach, Va. One of the four largest U.S. air shows, Oceana features the Blue Angels and a very long list of other flying performers. oceanaairshow.com

EDITOR'S CHOICE
» Reno Air Races
Sept. 13–17, Reno, Nev. Five days of racing by six classes of aircraft. Racers exceed 500 mph; includes the Air Force Thunderbirds. airrace.org

» Robodock
Sept. 20–23, Amsterdam This unique, 9-year-old international summer festival merges technology and art. robodock.org/home_eng.html

October

» Trinity Site Visit Day
Oct. 7, near Alamagordo, N.M. The birthplace of the atomic bomb, Trinity Site is open only twice a year. See where the nuclear era began. www.wsmr.army.mil/pao/TrinitySite/trinst.htm

» THEMIS Satellite Launch Oct. 19, Kennedy Space Center, Fla. A Delta 2 Rocket takes the "Time History of Events and Macroscale Interactions during Substorms" (THEMIS) spacecraft into orbit. kennedyspacecenter.com

» X Prize Cup
Oct. 19–22, Las Cruces, N.M. Astropreneurs try to launch their reusable rocket ships multiple times in under 24 hours. xpcup.com

» Morton Punkin Chuckin' Oct. 21–22, Morton, Ill. Calling itself "The Pumpkin Capital of the USA," Morton organizes one of the largest assemblages of pumpkin-hurling catapults and air cannons in the country. pumpkincapital.com

» Bridge Day
Oct. 21, Fayette County, W. Va. This may be the largest annual "extreme sports" event in the world. BASE jumpers parachute from a steel bridge in front of 200,000 spectators. bridgeday.info

» PhreakNIC X
Oct. 20–22, Nashville, Tenn. Originally started as a "hacker convention," PhreakNIC now includes sci-fi/fantasy, gaming, anime, and other areas of tech culture. phreaknic.info

» The Texas Mile
Oct. 28–29, Goliath, Texas Automobiles, motorcycles, and land speed racers push machines to speeds over 250 mph on a round on the runway of a remote Texas airport. texasmile.com

EDITOR'S CHOICE
» Edwards AFB Open House Oct. 28–29, Edwards Air Force Base, Calif. In addition to hosting one of America's premier air shows, the Edwards AFB open house allows access to the Air Force Flight Test Center and NASA's Dryden Flight Research Center. www.edwards.af.mil

» Baltimore Flugtag
Oct. 21, Baltimore, Md. Thirty teams take to the sky above the Inner Harbor for Flugtag, which means "flying day" in German. This event showcases the creative minds among us who build human-powered flying (read: falling) machines. They glide off the end of a 30-foot ramp hoping to achieve a few seconds of flight before landing in the harbor. redbullflugtagusa.com

November

EDITOR'S CHOICE
» World Championship Punkin Chunkin
Nov. 3–5, Millsboro, Del. Tens of thousands converge on a Delaware cornfield to watch air cannons, trebuchets, and other hurling devices vie to hurl a 10-pound pumpkin the greatest distance. punkinchunkin.com

» Sta-Bil Turf Classic
Nov. 4, Shelby, N.C. This final U.S. Lawn Mower Association sanctioned event features spirited and sort-of-serious lawn mower racers. letsmow.com

» Dawn Spacecraft Launch Nov. 11, Kennedy Space Center, Fla. A Delta 2 rocket launches the Dawn spacecraft on a scientific mission to asteroids Ceres and Vesta. kennedyspacecenter.com

Important: All times, dates, locations, and events are subject to change. Verify all information before making plans to attend.

Know an event that should be included? Send it to: events@makezine.com. *Sorry, it is not possible to list all submitted events in the magazine, but they will be listed online.*

If you attend one of these events please tell us about it at forums.makezine.com.

TALES FROM MAKE: BLOG

iTune In, Turn On, Make Stuff

By Phillip Torrone

Looking for a weekend project or steam-powered robots? All aboard! Take a tour of the new video features, including projects, interviews, and virtual tours.

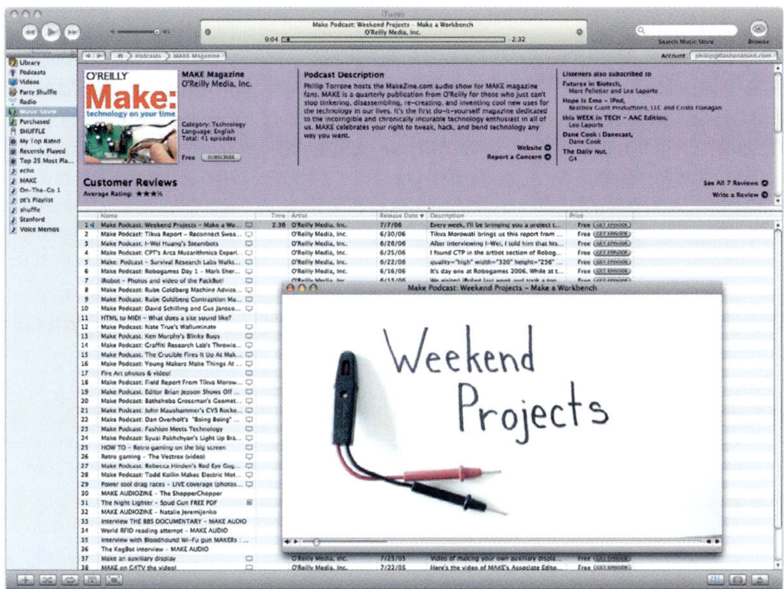

Get MAKE's new Weekend Projects how-to video podcasts delivered fresh through iTunes.

The internet is made out of tubes,

and now we're filling them with talkies, thanks to our new Media Maker, Bre Pettis. Formerly a puppet maker and schoolteacher, Bre is filming all things MAKE and serving up video on the MAKE: Blog (makezine.com/blog). Here's a tour:

Weekend Project

Every week, we're featuring a project that you can make over the weekend. For the first video podcast, you can learn **how to make a workbench** for your garage, studio, or get your priorities straight and put it in your living room! We also posted the Sketchup files; Sketchup is a free 3D design tool for creating ideas (sketchup.google.com). Download Sketchup and go look at the workbench in the Google Sketchup warehouse, or just download the Sketchup file on makezine.com.

Interviews

I-Wei Huang creates amazing robots that run on steam. An animator by day and steam-roboteer at night, I-Wei's inspiration for his robots includes animators such as Miyazaki (*see page 18 for an article on his work*).

David Schilling and Gus Jansson started up SMART, which stands for Seattle Mindstorms and Robotic Techies, and they've got a great community of Lego robotic enthusiasts. They work with other Lego robot makers in the Seattle area to create collaborative robotics.

Tours & More

Go where no maker has gone before and lived to tell of ... **a walkthrough of the Survival Research Labs.** SRL has been making industrial-sized remote control machines and performing with them around the world since 1978 (*see page 28 for a profile of SRL founder Mark Pauline*).

Then take a visit to **iRobot** and look at the evolution of their bots, from the Roomba robot vacuum (with its open interface) to the PackBot.

Get the Vids

Get the videos (MP4) delivered automatically with iTunes for playback on PC/Mac/Linux/PSP and iPod video devices (makezine.com/go/ivid), or grab them at YouTube (youtube.com/user/makemagazine) or Google Video (makezine.com/go/gvid).

Phillip Torrone is senior editor of MAKE.

Strange Love

Or, how they learned to start worrying and love to hate the bomb.

Physicists love explosions. We owe our nuclear predicament to a quirk of human nature: designing, making, and testing nuclear explosives can be fun. "The sin of the physicists at Los Alamos did not lie in their having built a lethal weapon," physicist Freeman Dyson (my father) has explained. "They did not just build the bomb. They enjoyed building it. They had the best time of their lives building it. That, I believe, is what Oppenheimer had in mind when he said that they had sinned."

Eight years ago, I began interviewing retired (and semi-retired) nuclear weaponeers who had worked on Project Orion — the technically promising but politically unacceptable effort, begun in 1957, to build an interplanetary spaceship propelled by nuclear bombs. The project's leader, physicist Theodore B. Taylor (1925-2004), exemplified the conflict between love of explosions and fear of the results.

"I was given a chemistry set when I was 7 or 8 and that rapidly turned into a laboratory for making explosives, with one restriction set down by my mother: never, never under any circumstances was I allowed to make nitroglycerine," said Taylor. "So I didn't." He experimented with more explosive and less stable alternatives instead. "I was fascinated by explosions. I still am. Without any attraction to the damage. I hated to just fiddle around. I wanted to go to extremes."

Taylor promised his mother, in the aftermath of Hiroshima, that he would never work on nuclear weapons, but the temptation proved impossible to resist. After an unsuccessful first attempt at a Ph.D., Taylor with his wife, Caro, and four-month-old Clare, drove their 1941 Buick to Los Alamos from Berkeley in November of 1949. "Within 24 hours of our arrival at Los Alamos, I was deeply immersed in the nuclear weapons program. Within a week, I was hooked on understanding what went on at these enormously high energy densities, clear off any human scale."

Within four years — but still without a Ph.D. — Taylor's designs included the largest, the smallest, and the most efficient fission devices ever exploded. The first of these records still stands. This was the Super Oralloy Bomb, which yielded 500 kilotons in the Ivy King test at Eniwetok on Nov. 15, 1952.

"I had complete freedom to work on any new weapon concept I chose," Taylor told me. "It's an exhilarating experience to look at what's going on inside something the size of a baseball that has the same amount of energy as a pile of high explosive as big as the White House. I went crazy over that. A big high. The highs needed fixes. And we got those twice a year easily. The fix was a combination of seeing one of these things go off — 'Aha! It worked!' — and seeing how the next one might be even more spectacular."

The first test witnessed firsthand by Taylor was Greenhouse Dog, at Eniwetok, yielding 81 kilotons on April 7, 1951. He was 15 miles away. "The explosion was every bit as awesome as I had expected — roughly five times as big as the one that destroyed Hiroshima. The countdown started close to dawn ... 1 minute ... 30 seconds (put on your dark goggles) ... 15 ... 4, 3, 2, 1: instant light, almost blinding through the goggles, and heat that persisted for a time that seemed interminable. The back of my neck felt hot from heat reflected off the beach house behind us. Goggles came off after a few seconds. The fireball was still glowing like a setting sun over a clear horizon, a purple and brown cloud rising so fast that in less than a minute we had to crane our necks to see the top. I had forgotten about the shock wave, a surprisingly sharp, loud crack that broke several martini glasses on the shelf of the beach house bar. I tried hard to shake off the feelings of exhilaration, and think about the deeper meanings of all this, without success."

The following year, at the Nevada Test Site, Taylor held up a small parabolic mirror and lit a cigarette with an atomic bomb. The fireball was 12 miles away. "I carefully extinguished the cigarette and saved it for a while in my desk drawer at Los Alamos," he remembered. "Sometime, probably in a state of excitement about some new kind of bomb, I must have smoked it by mistake."

What excited Taylor most were really, really small atomic bombs. "It was curiosity, wondering, 'What's the limit?' I wanted a panoramic view." Taylor was interested in low-yield explosions not because he anticipated a need for them — or a fear of terrorism — but because he was intrigued

Test Shot HOW, Operation TUMBLER-SNAPPER 3:55 a.m. June 5. 1952, Nevada Test Site, tower detonation at 300 feet, yield 14 kilotons. Photographed by an automatic ultra high-speed camera 0.0008 seconds after detonation. This was a developmental test of Ted Taylors's small, light, high-compression device that used a revolutionary beryllium tamper. Taylor lit his cigarette with the detonation's light focused in a parabolic mirror.

Photograph of HOW, 14 kilotons, Nevada, 1952, from *100 Suns*, by Michael Light, 2003

by the delicate balances involved.

"I said, why don't we build things with much less plutonium in there and see what's going on in the middle with much more sensitivity. We can do things at around a kiloton instead of what was then the predicted yield of a stockpile bomb, 80 kilotons — it was that for years. To make small yields with big implosion assemblies, that got fascinating. I was pushing things as far as one could go, never mind that you wind up in some cases with shells less than a millimeter thick. Who's going to make those? As it turned out, it was very worthwhile to find some way to make those."

"Pursuing these limits became an obsession," Taylor admitted. "What is the absolute lower limit to the total weight of a complete fission explosive? What is the smallest amount of plutonium or uranium 235 that can be made to explode? What is the smallest possible diameter of a nuclear weapon that could be fired out of a gun?" The answers were surprising. "I was narrowing my focus, getting the quantities of plutonium that one could use to make nuclear explosions down to less than a kilogram. Quite a bit less."

The smallest tactically deployed nuclear weapon was the Davy Crockett, with a warhead weighing less than 60 pounds. It was not designed by Taylor. "I tried to find out what was the smallest bomb you could produce, and it was a lot smaller than Davy Crockett, but it was never built in those years," he said. "It certainly has been since then. It was a full implosion bomb that you could hold in one hand that was about 6 inches in diameter."

Taylor left Los Alamos in 1956 to work for General Atomic, first on the TRIGA research reactor and then on Project Orion, and then left General Atomic to work for the Pentagon's Defense Atomic Support Agency in 1961. He was surprised to learn how much fissile material was lying around. He began to think about do-it-yourself nuclear weapons, and became alarmed.

"The use of small numbers of covertly delivered

The Davy Crockett, ready to be deployed. It had a warhead that was less than 12 inches in diameter, had a weight of about 60 pounds, and yielded up to a kiloton.

nuclear explosives by groups of people that are not clearly identified with a national government is more probable, in the near future, than the open use of nuclear weapons by a nation for military purposes," he warned in November of 1966, in his privately circulated "Notes on Criminal or Terrorist Uses of Nuclear Explosives." Keeping fissile material out of the hands of foreigners might not be enough. "The group could be an extremist group of U.S. citizens who believe they are trying to save the U.S."

Although Project Orion was conceived as a way of expending our stockpile of nuclear weapons to explore the solar system, Orion's physicists soon found that to gain the support of the nuclear establishment they had to answer the question: could they launch Orion without depleting the stockpile? Fortunately or unfortunately, the answer was yes.

"One of the big questions, a large part of the whole project which I cannot talk about freely, is just how much plutonium you need," Freeman Dyson explained in 1999. "One of the things that made Orion very attractive is the trade-off between plutonium and high explosive. In the ordinary bombs we use for the stockpile, all kinds, it doesn't matter whether they are high yield or low yield. The military likes minimum weight and minimum volume, so you tend to use a rather small amount of high explosive because it quickly becomes the dominating mass. For what we wanted to do, it was an advantage to have a huge amount of high explosive because that would also absorb neutrons and be the shielding for the ship."

He added, "Then you need a lot less plutonium. And how much less I cannot discuss. The whole economy of the thing depended on that. These were all very nonstandard bombs, which meant nobody believed us; the numbers clearly didn't add up. This is also an interesting question from the point of view of the terrorist bomb problem. If you have a bunch of people wanting to blow up the World Trade Center or something, they might have no difficulty getting large amounts of high explosive. So it is important not to declassify all that stuff."

Consequences of the laws of physics can only be concealed for so long. "Scientific secrets do not keep," warned Edward Teller, cautioning us to acknowledge that we can never maintain a monopoly on secrets such as how little fissile material is required to build a bomb.

There were four main technical obstacles to building an implosion weapon the first time: accumulating fissionable material; performing the computations necessary to validate the physics underlying the design; machining the components precisely in space; and firing the detonators precisely in time. Computers have shifted the landscape, and only the first obstacle still looms large. The average notebook computer has more computing horsepower than all of Los Alamos did while the weapons constituting most of our present stockpile were designed.

Since the first nuclear explosion on July 16, 1945, at Alamogordo, the growing club of nuclear powers have conducted approximately 2,000 nuclear tests: in the atmosphere, in space, underwater, and underground. Surely we are safer now that atmospheric testing has stopped? Maybe not. The risk from fallout has dropped. But we may owe the restraint that kept us away from the nuclear precipice over the past 60 years to nuclear policy makers who had actually seen bombs go off. All the weaponeers I interviewed, no matter how convinced of the need for overwhelming nuclear force as a deterrent, prefaced their statements by describing the effects of being an eyewitness to a nuclear test.

"I was there at the big one on Bikini," retired Air Force Col. Donald Prickett told me, over pancakes made from a sourdough culture he had nurtured uninterrupted for 54 years. This was Castle Bravo, exploded on Feb. 28, 1954, with a yield of 15 megatons, almost three times what had been

Photograph courtesy of U.S. Army, National Archives

expected, producing a fireball more than three miles across.

"I had seen up until that time maybe 50 shots at least, atmospheric shots out at the test site, so I wasn't really startled," said Prickett, describing how, with Navy Capt. George Malumphy, he maneuvered a remote-controlled merchant ship into the path of the fallout to test an automatic wash-down system being developed for decontamination of surface craft. "I knew it was going to be big, but Malumphy and I were at least 30 miles from ground zero. And so when the order came on for countdown, we put on our dark goggles. And sure enough it went off, and it was a full two minutes anyway before we took off our goggles, and then it was so awesome that all Malumphy could say was, 'My God, my God, my God!'"

Prickett, who died in 2004, wants us to remember what he could not forget. "I wish people could understand what would happen if one of these megatons ever got over to these cities. I wish to hell these people could see something like that. You're going to have to keep indoctrinating people to what these things are. Or they will forget."

"I wish to hell these people could see something like that. You're going to have to keep indoctrinating people to what these things are. Or they will forget."

On May 28, 1998, I was spending the day with Ted Taylor when news came in that Pakistan had conducted a series of nuclear tests. I expected a somber response. But Taylor was unable to conceal the old excitement: "Aha! It worked!" Over dinner, he kept drifting away from the conversation and coming back with some new insight, based on the sketchy news reports that had come in during the day, as to what his Pakistani colleagues had tested, and what they might do next.

Pakistan wanted to show the world (and India) that they had joined the nuclear club. Before the countdown, they disconnected all seismographs, not to conceal a successful test but to conceal their failure in the event the devices fizzled out.

The latest advance in the United States nuclear arsenal is the stockpile stewardship program, which claims to predict, purely from computer simulations and non-nuclear tests, whether our stockpile weapons will work or not. The next step in this arms race is a new generation of weapons whose designs are so simple, and so completely modeled using powerful computer simulations, that we do not have to test them to be sure that they will explode. But this favors potential adversaries as much as it favors us. The danger of not testing nuclear weapons is that we no longer know who has what.

"I had a dream last night, about a new form of nuclear weapon, and I'm not telling anybody what this is, because I'm really scared of it," Taylor told me in 1999. "I have tried, I thought successfully, to hold on to a vow of just not thinking about new types of nuclear weapons any more. And what's happened, to put it simply, is that it has gone from my conscious to my unconscious, and it's emerging as a dream; I cannot shut it off. I woke up at 2 a.m. and went back to bed at about 6 o'clock, and wound up filling up a page with notes. It makes me think of the prototypical example of what directed energy can do, making the transition from a pile of high explosive to a gun, as the Chinese did, after they invented it. What I am afraid is in the offing is people figuring out how to make a transition that's as spectacular as trying to kill a deer at 200 yards with a pile of high explosive, or by shooting at it."

Taylor had the time of his life designing bombs, and spent the remainder of it trying to get the madness of threatening to use them stopped. His final words to me: "I am searching for the truth as long as I can."

We are now relinquishing control of our nuclear arsenal, for the first time, to a generation that has never seen a nuclear explosion firsthand. There are no more Ted Taylors. The new generation of nuclear weaponeers grew up with video games, but was not allowed to have chemistry sets. Are we any safer as a result?

Further reading: *Curve of Blinding Energy* by John McPhee, and *Project Orion* by George Dyson.

George Dyson, a kayak designer and historian of technology, is also the author of *Baidarka* and *Darwin Among the Machines.*

Photograph by Dave Prochnow

HOMEBREW

My Robosapien in a Can

By Dave Prochnow

Harkening back to a day when youngsters called the local dime store and asked, "Do you have Prince Albert in a can?", I answered the call for entries on the Altoids website for the "Tin Million Uses" contest with my modern take on this well-worn prank — Robosapien in a Can.

Attempting to squeeze the 14-inch-tall Robosapien into the 1-inch-tall Altoids peppermint tin seemed simple enough. Heck, I'd been hacking the toy for over a year. So I was prepared to spend about an hour on this contest entry. *Wrong-o, bucko.*

While just slapping the Robosapien's main circuit board and speaker inside the tin would have been enough, I wanted to make sure this wasn't just some con job. My Robosapien in a Can would be fully functioning, with sensors, speech, and IR remote control.

I dispatched Robosapien's outer skin and extracted the main circuit board and all of the connected sensors in just under 40 minutes. *Ta-dah.*

It doesn't take much of a robot design genius to figure out that four D-size batteries ain't gonna fit inside an Altoids tin. But my Robosapien in a Can wasn't going to drive any servomotors, so a smaller power source would be acceptable. I substituted a 3V lithium coin cell for the original D-size power pack.

Now it was time to cram it all into the tin. The battery and speaker were placed side by side on the bottom. Next, the main circuit board was insulated with the tissue liner that comes packaged with each tin of Altoids. I mounted the power switch on the side of the tin, and coiled the microphone on top of the main circuit board. I was running out of room — fast.

Sacrifices had to be made. I removed the touch sensors from the main board, shortened the original wiring, drilled holes in the lid for the LED eye array, and fixed the IR receiver to the side of the tin. Most regrettably, a new latch had to be installed on the tin.

But it worked. Even scrunched down and fixed with my new copper latch, Robosapien in a Can worked. Did I win the Altoids contest? Nope. But I still had a wonderful homebrew Robosapien I could leave on my desk and scare the pants off anyone who tried to steal one of my curiously strong, high-tech mints.

Dave Prochnow is a frequent contributor to MAKE, *Nuts and Volts*, and *SERVO* magazines, as well as the author of 25 nonfiction books. Dave's website is at pco2go.com.